Becoming a Rockstar SRE

Electrify your site reliability engineering mindset to build reliable, resilient, and efficient systems

Jeremy Proffitt

Rod Anami

BIRMINGHAM—MUMBAI

Becoming a Rockstar SRE

Group Product Manager: Mohd Riyan Khan

Publishing Product Manager: Surbhi Suman

Senior Editor: Romy Dias

Technical Editor: Shruthi Shetty

Copy Editor: Safis Editing

Project Coordinator: Ashwin Kharwa

Proofreader: Safis Editing

Indexer: Tejal Daruwale Soni

Production Designer: Alishon Mendonca

Marketing Coordinator: Agnes D'souza

First published: March 2023

Production reference: 1290323

Published by Packt Publishing Ltd.

Livery Place

35 Livery Street

Birmingham

B3 2PB, UK.

ISBN 978-1-80323-922-4

www.packtpub.com

For my wonderful wife, who still likes me after 18 years. I like you too.

– Jeremy Proffitt

To my God, wife Tati, and son Gabe.

– Rod Anami

Contributors

About the authors

Jeremy Proffitt (born January 1977) is obsessed with constantly improving systems and solving problems with an unmatched sense of urgency – the definition of a **Site Reliability Engineer** (**SRE**). A master of solutions and technological knowledge, Jeremy is a rockstar SRE with AWS professional certifications in Architecture and DevOps – and has routinely saved millions in potential lost revenue in his career. In his free time, Jeremy enjoys spending time in his rockstar-appropriate technology cave and loves venturing into 3D printing, electronics, and **Internet of Things** (**IoT**) projects. By day, Jeremy currently manages a team of top SRE and DevOps talent driving constant improvement and is often cited in the company as a visionary in terms of observability and emergency response.

To the leaders who have helped me see the truth in our work and friends who have stood by and given me the encouragement to follow the wonders of technology, often while in awe of their own work, I say thank you! To my arch-enemies, you have been a wonderful addition that has always challenged me to become better. And finally, to my wife, Jamie, who I still desperately love after 18 years – and mind you, still likes me – I still remember our first date when you took my arm, you stole my heart, and in all our years, I've never felt you let go once.

Rod Anami is a seasoned engineer who works with cloud infrastructure and software engineering technologies. As one of the SREs at the Kyndryl CoE, he coaches other SREs on running IT modernization, transformation, and automation projects for clients worldwide. Rod leads the global SRE guild inside Kyndryl, where he helps plant and grow SRE chapters in many countries. Rod is certified as an SRE, technical specialist, and DevOps engineer professional at the ultimate level. He holds AWS, HashiCorp, Azure, and Kubernetes certifications, among many others. He is passionate about contributing to open source software at large with Node.js libraries.

I want to thank my wonderful wife, Tatiana, and my beloved son Gabriel, for giving me the space and support needed to write this book. My parents, Shizuo and Rita, for raising me with solid character. The Google site reliability engineering organization made this fantastic approach and profession open source. I want to thank Kyndryl for backing me on this journey. I had many bosses and leaders, good, bad, and inspiring ones. I want to mention a few who impacted my career immensely by helping me acquire the skills and knowledge for this book: Marcos Cimmino, Tara Sims, Andy Barnes, and Gene Brown. Nothing great is accomplished alone: it requires effort, endurance, enjoyment, colleagues, and God.

About the reviewers

Chris Smith is a strategic IT leader with a proven track record across the financial service industry. His passion is to lead organization-wide transformational efforts for Fortune 500 institutions within digital and contact center technology and operations. He is skilled at driving agile adoption, building an engineering-first mindset, and facilitating cloud modernization of core banking services at scale.

Itohanoghosa Eregie is the founder of techinanutshellhack, a platform dedicated to explaining technology concepts with short video clips about cloud and site SRE concepts in their simplest form via LinkedIn. She worked as a software developer at Cyberspace Limited before finding her passion as a platform engineer, which earned her an opportunity to work with Dell EMC as a resident platform engineer for one of Africa's largest telecommunications companies, MTN Nigeria, as a platform engineer. Altoros Americas currently employs her as a VMware Tanzu engineer, involved in customer engagement. Itohan is passionate about building resilient systems in the cloud and ensuring organizations adhere to SRE practices.

Brannen Taylor has almost 30 years of experience in corporate IT from the healthcare, managed services, power, hosted DR, and financial services industries. He has worked with small "mom-and-pop" operations up to ITIL-heavy Fortune 10 companies. He was a network engineer for 20 years and has been a network operations manager for the past 2 years. He has certifications from many vendors such as Nortel, Cisco, and Palo Alto, as well as a few that are vendor-agnostic, many cloud certifications from AWS and Azure, and is now moving into **Network DevOps** (**NetDevOps**), focusing on Nautobot, Ansible, and various vendor SDKs. He enjoys scuba diving with his wife and friends and has two grown children.

I would like to thank God for leading me into a career that I love. I want to thank my children for only eye-rolling me a little when I launch into an explanation about binary when they ask me how email works. I want to thank my wife Lara for putting up with me being on call these past 23 years, working unexpectedly long days, nights, and weekends, and non-stop studying. Thank you to my colleagues and the friends I've made along the way.

Gene Brown is the Vice President and a Distinguished Engineer at Kyndryl. He leads the SRE profession and certification program and is the global site reliability engineering leader. He is responsible for driving the enablement of SREs across Kyndryl's countries, practices, and strategic markets through a Center of Excellence with SRE chapter leaders across the services organization globally.

Gene enjoys spending time with clients interested in adopting SRE and likes comparing notes on what has worked well and how to overcome the challenges that come with cultural change. Gene was the co-founder of IBM's and Kyndryl's SRE profession with a focus on certifying SREs based on their applied experience in the field of site reliability engineering.

Table of Contents

Part 2 - Implementing Observability for Site Reliability Engineering

4

Essential Observability – Metrics, Events, Logs, and Traces (MELT)   57

5

Resolution Path – Master Troubleshooting   87

6

7

Part 3 - Applying Architecture for Reliability

8

Reliable Architecture – Systems Strategy and Design   165

9

Valued Automation – Toil Discovery and Elimination   189

10

Exposing Pipelines – GitOps and Testing Essentials   205

11

Worker Bees – Orchestrations of Serverless, Containers, and Kubernetes   227

12

Final Exam – Tests and Capacity Planning   247

Part 4 - Mastering the Outage Moments

13

First Thing – Runbooks and Low Noise Outage Notifications   267

14

Rapid Response – Outage Management Techniques   285

15

Postmortem Candor – Long-Term Resolution   303

Part 5 - Looking into Future Trends and Preparing for SRE Interviews

16

Chaos Injector – Advanced Systems Stability   319

17

Interview Advice – Hiring and Being Hired   341

Appendix A – The Site Reliability Engineer Manifesto   357

Appendix B – The 12-Factor App Questionnaire   359

Preface

Site reliability engineering relates to constant improvement, bridging business and product issues as per customer requirements and technology limitations, thereby generating higher revenue. Quantifying and understanding reliability, resource handling, and developer needs can sometimes be overwhelming. *Becoming a Rockstar SRE* explores reliability from an infrastructure and coding perspective and uses real-world examples to bring forth the **site reliability engineer (SRE)** persona.

This book will acquaint you with who an SRE is, followed by discussions on the why and how of site reliability engineering. It walks you through the jobs of an SRE, from automation of **continuous integration/continuous delivery (CI/CD)** pipelines and reducing toil to the details of reliability and the best practices to excel in it. You'll learn why harmful code is created and how to circumvent that with reliable designs and patterns. You'll explore how to interact and negotiate with businesses and vendors on various technical matters. You'll then deep dive into observability, outage, and why and how to craft an excellent runbook. Finally, you'll learn how to elevate your site reliability engineering career, including certifications, interview tips, and questions.

By the end of this book, you'll be able to identify and measure reliability, reduce downtime, troubleshoot outages, and enhance productivity to become a true rockstar SRE!

Who is this book for

This book is intended for IT professionals, from developers looking to advance into an SRE role to system administrators mastering technologies and executives experiencing repeated downtime in their organizations. This book will also be helpful to anyone interested in bringing reliability and automation to their organization to drive down customer impact and revenue loss while increasing development throughput. While reading this book, a basic understanding of API and web architecture and some experience with cloud computing and services will be helpful.

What this book covers

Chapter 1, *SRE Job Role – Activities and Responsibilities*, talks about the site reliability engineer persona addressing who is an SRE.

Chapter 2, *Fundamental Numbers – Reliability Statistics*, shows how the site reliability engineering work and business impact are measured.

Chapter 3, *Imperfect Habits – Duct Tape Architecture and Spaghetti Code*, explains why systems are naturally unreliable.

Chapter 4, Essential Observability – Metrics, Events, Logs, and Traces (MELT), discusses how we go from monitoring to true observability.

Chapter 5, Resolution Path – Master Troubleshooting, lectures on the SRE way of precisely and concisely troubleshooting.

Chapter 6, Operational Framework – Managing Infrastructure and Systems, describes why and how SREs tackle operational work and not just engineering duties.

Chapter 7, Data Consumed – Observability Data Science, teaches the basic mathematical models and statistical methods for SREs.

Chapter 8, Reliable Architecture – Systems Strategy and Design, describes systems thinking applied to reliability and reliable architectural patterns.

Chapter 9, Valued Automation – Toil Discovery and Elimination, familiarizes readers with a critical pillar of site reliability engineering: making operations scalable.

Chapter 10, Exposing Pipelines – GitOps and Testing Essentials, illustrates how to leverage reliability inside DevOps delivery pipelines.

Chapter 11, Worker Bees – Orchestrations of Serverless, Containers, and Kubernetes, presents how workload management affects the reliability of systems.

Chapter 12, Final Exam – Tests and Capacity Planning, demonstrates how good testing and capacity planning keep the performance of systems ahead.

Chapter 13, First Thing – Runbooks and Low Noise Outage Notifications, discusses how well-designed procedures and notifications prepare SREs for problems.

Chapter 14, Rapid Response – Outage Management Techniques, teaches about SRE positive behaviors and how to keep interactions toward the resolution during a significant incident.

Chapter 15, Postmortem Candor – Long-Term Resolution, portrays how postmortems should lead to actions that will make systems more reliable.

Chapter 16, Chaos Injector – Advanced Systems Stability, clarifies how SREs inject chaos into systems to learn more and use gamification to hone their skills.

Chapter 17, Interview Advice – Hiring and Being Hired, displays how companies should hire SREs and how SREs should demonstrate their knowledge during an interview.

Appendix A, The Site Reliability Engineer Manifesto, depicts the primary responsibilities of any SRE in the world.

Appendix B, The 12-Factor App Questionnaire, consolidates a series of questions to test whether an application design is reliable according to the twelve-factor app manifesto from Heroku.

To get the most out of this book

We purposefully used SRE as the acronym for site reliability engineer and kept site reliability engineering in its extended form throughout the book. For us, site reliability engineering is only accomplishable if you have an SRE and not the other way around. Although it's common to see SRE standing for both site reliability engineer and engineering interchangeably, we want to emphasize the persona and the who in this book.

This book contains simulation labs to give its readers practical knowledge. Each has a prerequisite knowledge set, such as Kubernetes, cloud computing, or software development. It's not part of this book to teach you about specific technologies and products but the most effective practices and principles that are technology agnostic. However, we must adopt some technology to demonstrate the site reliability engineering concepts and techniques. For that, we preferred open source software and platforms with free tier accounts in the labs.

Each simulation lab states its learning requirements and points to where the reader can find more information and instructions. We divided each practical exercise into three parts:

- Lab architecture
- Lab contents
- Lab instructions

The lab architecture explains the big picture around the design and connections among its main components. The contents section explains what's inside the GitHub repository, such as files and folders. And the lab instructions have a procedure for installing, configuring, and using the lab properly.

The following is a list of software covered in this book's simulation labs and the required execution environment:

Software covered in the book	Cloud platform requirements
GitHub	Google Cloud Platform account
Kubernetes	AWS account (alternative)
Node.js	Microsoft Azure account (alternative)
Prometheus	**Google Kubernetes Engine (GKE)**
Grafana	**Google Compute Engine (GCE)**
Terraform	Amazon Elastic Kubernetes Service (alternative)
Python	Azure Kubernetes Service (alternative)
Golang	
Slack	
PagerDuty	
Grype	

Syft	
Argo CD	
Grafana k6	
LitmusChaos	

You will require a laptop with reasonable access to the internet to work in the book's labs.

If you are using the digital version of this book, we advise you to type the code yourself or access the code from the book's GitHub repository (a link is available in the next section). Doing so will help you avoid any potential errors related to the copying and pasting of code.

Download the example code files

You can download the example code files for this book from GitHub at `https://github.com/PacktPublishing/Becoming-a-Rockstar-SRE`. If there's an update to the code, it will be updated in the GitHub repository.

We also have other code bundles from our rich catalog of books and videos available at `https://github.com/PacktPublishing/`. Check them out!

Download the color images

We also provide a PDF file that has color images of the screenshots and diagrams used in this book. You can download it here: `https://packt.link/W6q5Y`.

Conventions used

There are a number of text conventions used throughout this book.

`Code in text`: Indicates code words in text, database table names, folder names, filenames, file extensions, pathnames, dummy URLs, user input, and Twitter handles. Here is an example: "Within this repository, under the `Chapter-8` folder, there is just one subfolder called `terraform`."

A block of code is set as follows:

```
provider "google" {
  credentials = file("project-service-account-key.json")
  project = "autoscaling-simulation-lab"
  region  = "southamerica-east1"
  zone    = "southamerica-east1-a"
}
```

When we wish to draw your attention to a particular part of a code block, the relevant lines or items are set in bold:

```
resource "google_compute_autoscaler" "foobar" {
  ...

  autoscaling_policy {
    max_replicas    = 5
    min_replicas    = 1
    cooldown_period = 60
```

Any command-line input or output is written as follows:

```
$ terraform init
```

Bold: Indicates a new term, an important word, or words that you see onscreen. For instance, words in menus or dialog boxes appear in **bold**. Here is an example: "To do this, we navigate to the **Settings** tab in our GitHub repository. Then select **Secrets** on the left side, and choose **Actions**."

> Tips or important notes
> Appear like this.

Get in touch

Feedback from our readers is always welcome.

Join the SRE community: We invite you to join the large Site Reliability Engineers community at the `sreterminus.slack.com` Slack public workspace.

General feedback: If you have questions about any aspect of this book, email us at `customercare@packtpub.com` and mention the book title in the subject of your message.

Errata: Although we have taken every care to ensure the accuracy of our content, mistakes do happen. If you have found a mistake in this book, we would be grateful if you would report this to us. Please visit `www.packtpub.com/support/errata` and fill in the form.

Piracy: If you come across any illegal copies of our works in any form on the internet, we would be grateful if you would provide us with the location address or website name. Please contact us at `copyright@packt.com` with a link to the material.

If you are interested in becoming an author: If there is a topic that you have expertise in and you are interested in either writing or contributing to a book, please visit `authors.packtpub.com`.

Share your thoughts

Once you've read *Becoming a Rockstar SRE*, we'd love to hear your thoughts! Scan the QR code below to go straight to the Amazon review page for this book and share your feedback.

https://packt.link/r/1803239220

Your review is important to us and the tech community and will help us make sure we're delivering excellent quality content.

Download a free PDF copy of this book

Thanks for purchasing this book!

Do you like to read on the go but are unable to carry your print books everywhere?

Is your eBook purchase not compatible with the device of your choice?

Don't worry, now with every Packt book you get a DRM-free PDF version of that book at no cost.

Read anywhere, any place, on any device. Search, copy, and paste code from your favorite technical books directly into your application.

The perks don't stop there, you can get exclusive access to discounts, newsletters, and great free content in your inbox daily

Follow these simple steps to get the benefits:

1. Scan the QR code or visit the link below

https://packt.link/free-ebook/9781803239224

2. Submit your proof of purchase

3. That's it! We'll send your free PDF and other benefits to your email directly

Part 1 -
Understanding the Basics
of Who, What, and Why

In this first part, you will learn about site reliability engineering, its roots, and current usage outside Google. We emphasize how the **site reliability engineer** (**SRE**) persona is the center of gravity of everything orbiting systems reliability. When we talk about site reliability engineering, it's impossible to do so without a discussion about the business of software development, which we tie into not only statistics used for reliability but how those impact what companies are ultimately interested in, customer satisfaction and revenue. Finally, we'll explore why the lack of reliability persists in organizations and discuss some of the lesser known truths that make site reliability engineering critical and complex.

The following chapters will be covered in this section:

- *Chapter 1, SRE Job Role – Activities and Responsibilities*
- *Chapter 2, Fundamental Numbers – Reliability Statistics*
- *Chapter 3, Imperfect Habits – Duct Tape Architecture and Spaghetti Code*

1

SRE Job Role – Activities and Responsibilities

A lot has been said about **site reliability engineering**, what it is, what it is not, and the multiple practices and techniques that we should apply to adopt the site reliability engineering model. Who **site reliability engineers** (**SREs**) are is often put aside even though it is a crucial aspect. Moreover, how people from various parts of **information technology** (**IT**) become SREs and how some of them are recognized as thought leaders in this domain.

However, little has been said about the site reliability engineer persona, as detailed in the following list:

- What do they know?
- Which skills have they developed?
- What do they do daily?
- What are their primary responsibilities?

Those characteristics would explain, at a bare minimum, why someone should start the journey to becoming an SRE rockstar. That's precisely why we decided to start this book by outlining the SRE job role.

In this chapter, we're going to cover the following main topics:

- Making this journey personal
- Understanding the mindset and hobbies of an SRE
- DevOps engineers versus SRE versus others
- Describing an SRE's main responsibilities
- An overview of the daily activities of an SRE
- People that inspire

Making this journey personal

Unfortunately, often when an enterprise starts to adopt SRE into their IT governance processes, they don't use a **people-processes-tools** (**PPT**) model to transform their operations and software development areas, having a clear vision of these pillars. Even more often, they don't emphasize or focus on the **people** element of PPT in such transformations. We want to change that by making this learning journey personal and centered on the individuals rather than the involved processes or technologies.

It's critical to understand (and learn) what drives typical SREs forward, which fundamental skills they have developed, and how they hone their skills over time to go above and beyond at work. For that purpose, we will divide this subject into three sections:

- SRE driving forces
- SRE skills
- SRE traits

Let's start this personal journey by understanding why you should become an SRE.

SRE driving forces

We want to explore what motivates or incentivizes site reliability engineers. There's no journey of any nature if there is no driving force pushing you through. As a word of advice, we should warn you that learning about site reliability engineering is more of an expedition than a tourism trip. In other words, it's more a marathon than a sprint. Having clarified that, we'll begin by putting the possible rewards of this journey on the table. Let's depict each driving force as a mockup code snippet (**JavaScript**) to make it fun.

Money

If we could represent in the form of an algorithm how **money** drives people when they don't earn enough, it would look like the following:

```javascript
// money
if (money < MyMinimumSalary) {
motivated = false;
excitement--;
}
doMyWork();
if (motivated && jobSatisfaction) {
    honeSRESkills();
    doExtraWork();
} else lookForAnotherJob();
```

Site reliability engineers make more money than most other technical professionals. According to a *Glassdoor* (2022) report, they can earn more than USD 118K per year on average. In similar reports, SREs are even noted to have surpassed DevOps engineers in a salary comparison. Nevertheless, not making enough money can be a key demotivating factor. It is hard for anyone to move forward with their career if they are preoccupied with expenses.

Although SREs have a notorious income on average, their salaries will vary per country, years of experience, and employer. Companies justify SRE salary levels based on the reliability value they bring to the table. Rest assured, the site reliability engineering career is well paved in the compensation field.

Job satisfaction

What affects our **job satisfaction** can be depicted as code logic as follows:

```
// jobSatisfaction
if (interestingJob || purposefulWorkActivities ||
challengingSkillDevelopment || technicalAppreciation) {
    jobSatisfaction = true;
    excitement++;
}
```

Job satisfaction is another driving force of site reliability engineers, and it has many factors. We usually translate job satisfaction to employee happiness at work. Site reliability engineering leads to job satisfaction when we look at the following profession characteristics: exciting job content, purposeful work activities, challenging skill development, and technical appreciation.

The job content of site reliability engineering spans multiple domains. You can work with developers one day and help systems administrators the next. You may need to assist in redesigning an app to increase its service reliability. As with any generalist model job with technical depth in many subject areas, you will never get bored for sure.

As we will see later in this chapter, SRE work activities have clear business value. They improve not just the service quality, availability, and resiliency, but also the system's reliability. Reliable services might help with customer loyalty, bringing additional revenue to the service provider. There is a direct relationship between SRE work and business metrics improvement, making their efforts purposeful.

Since site reliability engineering is a cross-technology domain engineering discipline, any skills acquisition is challenging. SREs have knowledge and skills that a systems administrator or software developer doesn't have. They are required to keep those skills updated and hone them over time. This necessity to keep learning brings the always-moving-forward feeling that may not happen if you only need to master a single product or technology.

The last factor on our list is technical appreciation. According to **Boston Consulting Group (BCG)** research, appreciation is the number one job happiness factor. Being an SRE, you will aid customers, users, and other technical professionals because of your keen holistic view of the systems. Consequently, technical appreciation for the job you do is common, and who doesn't like that?

Innovative solutions

The following code gives you an idea of how exciting exploring uncharted terrains is:

```
If (!solutionExists) {
    deviseNewSolution();
    excitement++;
}
```

Site reliability engineers are natural trailblazers as they explore new technologies and processes to obtain better reliability and eliminate *toil* (manual and repetitive tasks that are devoid of value). They face many scenarios and situations that are a first of their kind. Moreover, they are responsible for paving the path for others by documenting procedures in runbooks when none exist. There's nothing more exciting than devising new solutions or improving existing ones. Imagine how you would feel if they named a technical operating procedure after you.

Nevertheless, SREs want to minimize complexity and reduce technical debt. They don't create a solution just for the sake of doing it unless it adds value and resolves or prevents events that impact customers.

Good relationships

The following code snippet is a representation of how **good relationships** are a result of an exciting working environment:

```
If (excitement > HIGH) {
    motivateOthers();
    relationships.healthy = true;
}
```

Also, good work environment relationships are one of the top 10 factors contributing to employee happiness. SREs have good relationships in their work environment. The reason is straightforward; they act as integration hubs among different tribes and have the mission to break company siloes. SREs need cooperation from both development and operations teams. They are technical diplomats and have strong communication skills. Since they are usually excited about their work, they tend to socialize more with colleagues and leaders, potentially helping to improve the social environment around them. That doesn't mean they need to be extroverts with progressive public-speaking skills, but certainly, SREs are good teachers because they are excited and compelled to talk about what they do.

SRE skills

Now that you know what's in it for you, it's time to check which skillsets SREs must develop throughout their careers. Site reliability engineers have a good mix of knowledge, skills, and experience that are shared with other roles and those that are unique to them. SREs have technical skills that span the entire solution life cycle, from the design to the manage step.

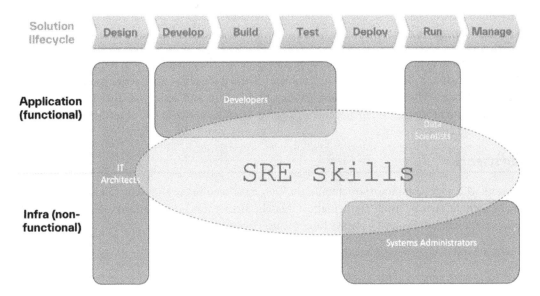

Figure 1.1 – SRE skills

The preceding Venn diagram shows how SREs acquire skills common to other professions and how SRE skills connect the various steps of the solution life cycle. In essence, site reliability engineers are senior technical resources that follow a generalist proficiency model with good depth at certain areas of expertise.

There's no consensus in the market about the canonical set of skills for SREs. It would not make sense for this to be the case because as soon as any technology-based skill becomes obsolete, we would need to remodel the whole profession. Instead, SRE core skills should be as technology-agnostic as possible.

We recommend a blend of distinct expertise from the IT architect, software developer, data scientist, DevOps engineer, and systems administrator roles. The proportion of each skill level varies per a multitude of factors. You will need to determine which skills are more in demand than the others, but an organization should have all of them in its toolkit.

Systems thinking

Site reliability engineers have a holistic view of the system's reliability by understanding the availability, resiliency, and performance of each solution component at both the application and infrastructure levels.

Software engineering

SREs develop code and software. They know how to utilize algorithms and software development techniques such as **agile** frameworks. SREs need to be proficient enough in instrumenting the app code to increase its manageability and observability. SREs know how to use software development life cycle tools and technologies, including DevOps (**continuous integration/continuous delivery – CI/CD**) pipelines. They can provide testers with better test cases that consider service reliability targets.

Systems management

SREs know how to manage, administer, and operate systems. They share most of the skills from the systems administrator role on multiple technologies. Their technology knowledge spans the cloud, containers, storage, networking, operating systems, middleware, and databases. They have the skills to implement monitoring, event management, logging, tracing, service levels, observability, DevOps toolchains, and automation of *toil*.

Data science

SREs work with huge amounts of structured data. They must acquire the knowledge and skills to make sense of such datasets by using mathematical models. SREs know how to analyze data to uncover trends, anomalies, and insights – always from the user's perspective.

We recommend that every SRE has the following selection of core skills:

- Systems thinking for focusing on the reliability of the system
- The ability to develop and test software
- The ability to deploy and release apps
- IT service management
- Systems monitoring and observability
- Working with DevOps tools and automation
- The application of data science for reliability of systems

Although we didn't explicitly mention **security** in any of the knowledge domains, enforcing security across multiple layers is present in all of them.

> **Important note**
> All fundamental SRE skills are covered in this book's chapters. The chapters have been organized to optimize your learning journey, so they don't follow the preceding order of skills. We structured this book based on our own experience acquired from a multitude of site reliability engineering coaching and mentoring sessions.

We provide a manifesto model in the *Appendix A, The Site Reliability Engineer Manifesto*, that acts as a more structured guide for site reliability engineering adoption, including the fundamental skills. We hope that helps your company in joining the site reliability engineering movement.

SRE traits

Besides what the SREs know and which skills they must develop, it's relevant to know their other good traits.

Software is everywhere

Site reliability engineers have a software engineering mindset. The idea of approaching any issue as a software problem may be disruptive at first; however, there is a good reason for it. Imagine that you need to restart and verify a system by manually issuing a specific set of commands and parameters many times per week. If you handle it as a software development problem, the solution will be developing and scheduling a simple program or automation to execute this task instead. SREs embrace automation over *toil* as one of their best tools.

Comfortable to code

SREs are not just able to develop code; they really like doing it. As we will see later in this chapter, they develop code as a frequent activity and main responsibility. It's not just a question of learning how to code or program when someone asks; SREs always feel confident in constructing good pieces of software.

Change as a constant

Frequent releases of new features, code enhancements, and reliability improvements are vital for any business. SREs are the first to accept calculated risks to provide more value to the system users. They are not risk averse but bring visibility to inherent risks so they can throttle the speed of change. They are always prepared to make progress and go above and beyond for service reliability.

Handle complexity and scale

They are not afraid of complexity or scale. They know that modern workloads are intrinsically complex and must be scalable horizontally and vertically. SREs work with large systems with multiple components running in hybrid multi-cloud environments. They understand the application's full-stack design, its moving parts, and how they connect to each other.

Problems as opportunities

SREs participate in *on-call* rotations and schedules to respond to service disruptions. They see incidents and problems as opportunities to learn and advance the reliability of the system. Not just that, they also have the competence to translate technology into business language to measure the impact on users and customers. They advocate for a *blameless* culture by prioritizing answers to questions, such as how to detect and repair incidents faster next time. They also consider how technical challenges may affect business results.

We have just gone over what makes the SRE persona: their motivations, skills, and traits. Now we are going to understand how site reliability engineers think.

Understanding the mindset and hobbies of an SRE

It's not rare for site reliability engineers to have a broader and divergent view of their surroundings. We are not saying that SREs are weird; well, they are in a certain sense, as they employ a relentless search for improving reliability in all things. However, we are referring to their *mindset* and how they approach the world.

In this section, we will explore different aspects of their thought process in the work environment and what they like to do in the job and outside it. We have divided this topic into three sections:

- SRE affinity game
- SRE guiding principles
- SRE hobbies

You may have asked whether site reliability engineering is the right profession for you. Let's examine that next.

SRE affinity game

Let's play a game! What do you think your affinity or compatibility is with the site reliability engineering profession? We will present a series of scenarios that SREs face. You need to answer them with either **love**, **like**, **dislike**, or **hate** indicating how much you see yourself doing it and how you would feel about it. Try to be as honest as possible.

> **Disclaimer**
>
> This is not an anthropological scientific survey based on a human behavioral model or theory by any means. It's a simple questionnaire to help you understand your own affinity to the SRE job role.

The scenarios are in the following list. Get a piece of paper, write down the question number, and answer it. Good luck!

1. Your boss asks you to resolve a problem that no one else has ever resolved.

2. You need to spend a few hours looking through logs, metrics, graphs, and events to verify whether there are any new anomalies that were not detected automatically.

3. You need to participate in an *on-call* rotation or schedule where you might be called late in the night to respond to a service disruption that has a business impact.

4. You need to work on a backend system or software that is not visible to external users.

5. You need to devise new ways to increase a large system's overall reliability.

6. You are asked to work on a large-scale problem, which affects hundreds of users and has dozens of components and dependencies, that runs on a hybrid multi-cloud environment.

7. You are diagnosing a system problem that is making users from a certain geography unable to access their services, and there is great pressure on you.

8. You need to approach problems with a selected scientific method or data model to uncover facts instead of guessing.

9. You constantly ask yourself how you could make things around you better and more reliable.

10. You need to classify and categorize systems information and functionalities so you can isolate causes from effects.

11. You must diagnose and fix a system problem by investigating components that are not usually visible by going deep into each component configuration as debugging mode is not available.

12. You need to design a detailed diagram of how the user interacts with a system or software so you can point out where to observe for symptoms.

After you complete this exercise, assign points to each of the answers. If you replied to a scheme with a **love** answer, assign 5 points to it. For a **like** answer, you get 3 points. **Dislike** has a value of **0**, and **hate** is **-3** (negative!). Sum your points across all 12 scenarios to get your score, and check the result against the following list:

- **Over 34 points**: Your affinity is very high; this is the right career for you

- **From 21 to 34 points**: Your affinity is high; you should consider this profession

- **From 13 to 20 points**: Your affinity is medium; this may be a good job role for you

- **Below 13 points**: SRE may not be your best option

This may be a game, but it will have made you imagine yourself in an SRE's shoes. We have started to understand the SRE mindset, so let's check what guides them in the convoluted scenarios listed previously.

SRE guiding principles

Everyone has a conjunction of principles (and values) that acts as their compass. SREs also follow a set of values; they embrace guiding principles to advise them on technical decisions and act as a reliability *compass*.

Google® coined most of those principles in its site reliability engineering books (`https://sre.google/books/`), but others appeared later in conference sessions at SREcon (`https://www.usenix.org/srecon`) and blog posts on many websites.

Again, we have selected some of them as canonical guiding principles based on our experience in assisting customers and organizations in enabling site reliability engineering in their IT shops. The following is the set of guiding principles that are rooted in the SRE persona:

- Scalable operations
- Engineering fidelity
- Observability to the core
- Well-designed service levels
- User-perspective notification trigger
- Blameless postmortems
- Simplicity

We must remark that such principles are not procedures or prescriptive instructions to accomplish something but guidelines. Don't worry if you are not familiar with the terminology applied here; we dig into them in a detailed manner throughout the book. Let's investigate each of them along with their most familiar patterns and anti-patterns.

Scalable operations

The operations team, which includes site reliability engineers, is responsible for managing production systems. They are the first responders for any service disruption when something goes wrong. The **scalable operations** principle states that this team will not grow proportionally to the system as its load increases. Another way to say that is if the number of active users for the determined service doubles, the operations team size will not double. A more mathematically accurate way to visualize this is through a logarithm growth curve. As the operations team gains technical maturity, eliminates repetitive manual tasks, and adopts automation at large, they will need fewer resources to manage more system load:

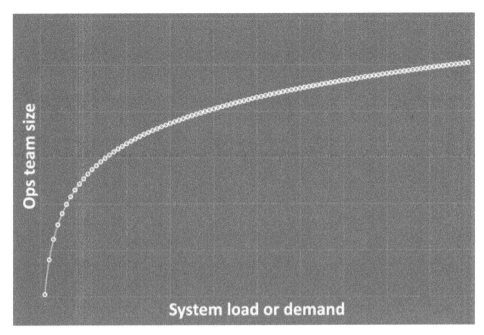

Figure 1.2 – A logarithm growth curve

It is worth mentioning that SREs employ a proactive approach as they strive to identify the root cause of issues and devise solutions to detect or prevent problems. The patterns for this principle are as follows:

- Identify and eliminate toil whenever possible

- Document operational procedures as runbooks

- Train operations teams to use and refine runbooks

- Adopt automation platforms and automated procedures documented in runbooks at large

The anti-patterns are as follows:

- Have linear (or exponential) growth for operations teams when the system load rises

- Operational knowledge is tacit or not documented

- Automation is the end goal and not merely a way to eliminate toil

Engineering fidelity

This tenet asserts the obvious: site reliability engineers do engineering. Yet it's not uncommon to see SREs only working on incident, problem, and change management processes. We are not telling you that site reliability engineers don't get their hands dirty; on the contrary, they do operational and engineering work. This principle exists to guarantee that SREs will have time to excel in both.

The patterns are as follows:

- Cap operational work at 50% of the available SRE time. The other half is dedicated to engineering solutions and increasing reliability.

- Share some of the operational work with the development team. Sharing 5% of the operational work is usually recommended, so the development team is prepared to take on SRE work.

- Send operational overflow work to the development team as they share the same goals.

The anti-patterns are as follows:

- SREs only work on operational work, resolving incidents, implementing changes, and running **root cause analysis (RCA)**

- SREs spend most of their time doing firefighting (incident resolution)

- Development teams don't share any responsibilities with the operations team

Observability to the core

Observability is the ability to comprehend the internal state of a system by inspecting its outputs. It extends the monitoring concept by adding layers to expand the system visibility and allows a more proactive posture by detecting anomalies before they become disruptions. This guiding principle craves visibility and discernability of what's happening inside a system or application by measuring certain signals.

The patterns of observability are as follows:

- Observe the system behavior through the **golden signals**; this can be either four (LETS) or five (STELA) signals depending on the school of thought you follow. The **LETS** acronym stands for **latency, error rate, traffic, and saturation**. **STELA** stands for **saturation, traffic, error rate, latency, and availability**.

- Have monitoring **metrics, events, logs, and traces (MELT)** at the SRE's disposal. These are the fundamental data components of any observability platform.

- Run synthetic user testing from time to time. This is a method where a bot mimics a user to test system functionality and response times.

The anti-patterns are as follows:

- Observe only the liveness of the system components, but not from the user's perspective. For example, checking that components are running versus checking that users can use the system as designed.

- Lack of user experience monitoring. You don't have visibility of what's happening in the user interface.

Well-designed service levels

There's no way to verify whether a service is being delivered to the target user within the expected and agreed-to parameters without established service levels. Part of the undeniable success of site reliability engineering is due to this redefinition of what a good service level is and how we document it. This tenet aims to have not just well-defined service levels but also well-designed ones that measure the system's reliability.

The patterns are as follows:

- Define **service-level indicators** (**SLIs**) from the system user angle, then delineate **service-level objectives** (**SLOs**) as an aggregation of the former
- Set the SLO target to less than 100%, so there's some room for errors (**error budget**) between 100% and the SLO target to launch new features and enhance overall reliability
- Establish **service-level agreements** (**SLAs**), with penalties and fines if they are not met after the measured SLOs
- Improve SLOs and increase their targets over time through engineering work carried out by the site reliability engineering team

The anti-patterns are as follows:

- Define the SLAs first, then measure the SLOs to see whether they are feasible with the current workings of the system.
- Establish a target of 100% for the SLAs or SLOs. This anti-pattern reduces the team's ability to release new features (or develop system reliability further) as there's no space for testing them in production. Soon enough, the whole system will become obsolete or non-competitive in its market.

User-perspective notification trigger

Notification is the process of alerting on-call first responders about service or system performance deterioration or downtime. It translates to when a site reliability engineer must be engaged to resolve an incident. This principle states that triggering a notification of an issue should only happen when this issue is affecting the system user. For instance, we never alert an SRE if the CPU load is high, but the user is not *feeling* any service degradation.

The pattern is as follows:

- Alerts are triggered if there are any symptoms at the user level, and if such warnings are actionable, SREs can resolve them

The anti-patterns are as follows:

- Alerting noise. SREs cannot differentiate between alerts that are mere informative events and ones that affect the system user.
- Lack of alerting. End users engage with the help desk to notify them that there are problems in the system.

Blameless postmortems

Postmortems are in essence **root cause analysis** (**RCA**) acts. They receive a peculiar name to avoid running into the same old pitfall: finding someone or something to be blamed (the root cause) rather than improving the system's quality and learning from mistakes. Postmortems also focus on questions, such as how to detect, respond to, and repair disruption in the service faster than just uncovering the root cause alone. This tenet is one of the hardest to deploy for a new organization if it has been doing traditional RCAs for some time and requires a blameless culture to support it.

The pattern is as follows:

- Infinite hows. Ask multiple questions, starting with the term *how*, to determine enhancements to the system (infrastructure and applications), processes, and knowledge base.

The anti-pattern is as follows:

- Go back to traditional RCAs where no progress is made on reliability

Simplicity

This guiding principle was imported from the **Agile Manifesto**. We can't explain it better than the manifesto (`https://agilemanifesto.org/principles.html`): "*the art of maximizing the amount of work not done.*" In other words, it dictates that site reliability engineers are always looking for ways of simplifying and avoiding unnecessary work. They are eager to eliminate *toil*, that is, repetitive, manually intensive, or low or no business-value tasks. However, inherently as humans, we tend to complicate everything, so ensuring runbooks are kept easy to observe and readable is a good example of this principle.

The pattern is as follows:

- **Keep it simple, stupid** (**KISS**) is a proven design principle from the US Navy that says most systems work better if they are simple to use or follow

The anti-pattern is as follows:

- Too elaborate processes for SRE work

We just explained our preferred seven guiding principles that site reliability engineers follow in their profession. They are an integral chunk of the SRE mindset. Let's now cover what SREs do in their free time to overcome learning limitations.

SRE hobbies

Jeremy and I couldn't agree more about what makes a site reliability engineer rockstar: their hobbies. What you do in your free time for leisure or as a second profession leads to greater levels of conceptual and practical knowledge. The trick is finding a hobby that you have a passion for and that helps in the SRE role.

We can't tell you what the best-fit extra-curricular activity that will pump up your SRE skills is, but here we list some examples that may interest you, grouped by the skills that they enhance.

Analytical thinking

Site reliability engineers have a good analytical processing capacity. They need to analyze big amounts of data and detect patterns, trends, and anomalies by correlating different data sources. Some engaging hobbies that leverage your analytical thinking are as follows:

- **Chess**: Without saying too much about it, this game has its own set of theories and algorithms. It is a good way to practice thinking multiple steps ahead while focusing on the present.
- **Board games**: There are plenty of board games that make you analyze information to win. And they make it enjoyable and social.
- **Rubik's cube**: This fun toy is also a good example of simplicity in operation and shape. It presents a complex challenge with a plain design.
- **Video games**: Strategy and role-playing games will train your mind in thinking analytically.

Creativity

SREs need to forge new algorithms for observability. They also need to construct new ways of measuring system reliability, as applications and infrastructure components have an uncountable number of arrangements, technologies, and architectures. This may sound cliché, but thinking outside the box is where site reliability engineers shine. Here are some hobbies that may help with your creativity:

- **Algorithms development**: Although this may be part of your daily work life, you can find fun in it by developing 2D or 3D video games, for instance. Another option is to contribute to open source software projects in the wider community.
- **Drawing or painting**: This is a relaxing and artistic example. It also gets you used to finding inspiration.

- **LEGO®**: An across-the-globe famous construction toy, LEGO makes you think about new forms, shapes, structures, and ways of assembly. It also has a robotics range that gives you programming skills as well.

- **Internet-of-Things prototyping**: How about developing embedded projects with Arduino® or Raspberry Pi® boards? You need to build both the hardware and firmware for the project to come together.

- **Video games**: The ones where you need to build something with blocks and basic structures, such as Minecraft®, are especially useful in nurturing creativity.

- **3D printing**: Author Jeremy is a 3D printing master. Like painting, you can express your art in three dimensions.

Troubleshooting

Troubleshooting is not exclusive to site reliability engineers. Systems administrators must also figure out why a service or system is down and how to repair it. However, SREs use systems thinking and scientific approaches to troubleshoot differently. You need to train your mind to resolve problems logically and calmly, and you're going to need it. There are plenty of hobbies that can stimulate you to excel in this area. Let's list some of them:

- **Crossword or jigsaw puzzles**: People are addicted to this type of entertainment. It's an excellent choice to keep the mind sharp and trained

- **Sudoku**: This was a trend not long ago, but it is still an excellent way to polish up troubleshooting skills.

- **Video games**: The ones full of puzzles, such as Portal, as especially good for troubleshooting practice.

We have now covered the aspects of the site reliability engineer persona. Next, we will look at what makes site reliability engineering professionals unique by comparing them to other roles and listing responsibilities and activities.

DevOps engineers versus SRE versus others

This is one of the most frequently asked questions we receive from customers and organizations: how does the site reliability engineering profession differ from other existing technical roles? We already talked about how SREs are the connection between the different steps of the solution life cycle. Here, we'll focus our discussion on the DevOps engineer role, and later, we'll broaden it. We have split this discussion into two sections:

- DevOps and site reliability engineers
- Software and site reliability engineers

DevOps and site reliability engineers

Google described the relationship between DevOps and SRE with a famous subtitle in their *The Site Reliability Workbook* publication:

Class SRE implements interface DevOps

This statement is an elegant way to define this link and refers to **Java** programming. It implies that site reliability engineering describes and deepens the implementation of whatever DevOps is. Moreover, we can say that site reliability engineering has commonalities with DevOps as a logically derived conclusion. However, what exactly does site reliability engineering implement from DevOps, or what are the differences between a site reliability engineer and a DevOps engineer? We have visualized these similarities and divergences in an infographic as follows:

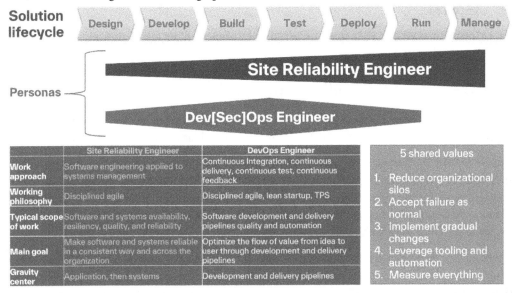

Figure 1.3 – An infographic on SRE and DevOps

Notice that they have shared values. Both SREs and DevOps engineers require those values in the orange (bottom right in the above diagram) box. In the bottom-left table, you can see the difference between those roles. Typically, site reliability engineers resolve operational problems by applying the right software engineering disciplines. On the other hand, DevOps engineers resolve development and delivery pipeline issues with systems management techniques mainly by using automation and infrastructure-as-code. They also concentrate different levels of effort on distinct phases of the solution life cycle, as depicted in the infographic.

It's not rare to hear that DevOps is a *shift-right* transformation while site reliability engineering is a *shift-left* one. That implies moving from the left (development side of the equation) to the right (operations side of the equation), and vice versa. Another term we hear a lot is **DevSecOps**, which has the addition of security. Since security has always been implied in these roles, we think including new letters in the middle is confusing and redundant.

SREs and DevOps engineers are, in our opinion, different sides of the same coin. They should be more like best friends forever than opposing roles as they share values. Let's check how SREs fulfill those values from the five main areas of DevOps:

- **Reduce organizational silos**: SREs use the same tooling as developers or DevOps engineers. They also share objectives and performance metrics with them.

- **Accept failure as normal**: SREs embrace risks using the *error budget* for new features. They quantify failure through SLIs and SLOs. And they run postmortems in a blameless culture.

- **Implement gradual changes**: SREs work to increase reliability, and more reliable systems allow more frequent changes and releases.

- **Leverage tooling and automation**: SREs eliminate *toil* by automating operational tasks at a constant pace.

- **Measure everything**: SREs measure reliability by implementing MELT data and observability. They also have ways to identify and size *toil*.

Software and site reliability engineers

Another frequently asked question is how site reliability engineers differ from **software engineers** (**SWEs**). The short answer is simple: they have the same core skills but specific work scopes.

What are **SWEs**? SWEs design, engineer, and architect applications using modeling languages and requirements analysis techniques. They implement an **integrated development environment** (**IDE**) and develop code for use cases using one of the multiple available programming languages. They create test cases and testing suites. Also, they integrate software and service components and handle their dependencies. SWEs work with many software development life cycle tools and processes.

Site reliability engineers may execute the same activities, but they intend to improve reliability when doing so. For instance, developing code for an SRE translates much more to instrumenting the application code, so it generates more logs, than coding a use case. Also, SREs treat operations as a software problem and see daily systems management tasks as possible software coding opportunities. Besides that, SREs have other core skills, relating to systems thinking, systems management, and data science.

Indeed, an SRE could become an SWE and vice versa, and that leads us to another principle that we find in the Google materials.

Common staffing pool

Another principle is hiring site reliability engineers and SWEs from the same staffing pool. This principle works well for companies where most employees are software developers and engineers, and having a shared pool means that site reliability and software engineering job roles are interchangeable. However, this principle may be much more challenging for enterprises with a mix of systems administrators and developers. Hence, we left it out of our list in the previous section.

We could compare the SRE's unique profession to many others, but we limited this topic to the most common comparisons. SREs are not architects, developers, systems administrators, or data scientists; they are more than all of these roles combined. Up next, we are going to understand the primary responsibilities of an SRE.

Describing an SRE's main responsibilities

We hope the SRE job role mission and scope are less foggy at this point. As an SRE, what would you be responsible for? In this section, we will investigate the most trivial duties that SREs are accountable for. We've divided these responsibilities into two sections:

- Operational work responsibilities
- Engineering work responsibilities

Let's start by reviewing the operational group first.

Operational work responsibilities

Site reliability engineers have work duties related to the process of managing systems. Such tasks are called operational work. SREs are not just accountable for operational work together with the operations team, but they also have the authority to execute their management processes.

First, they are responsible for the ITIL® processes, including incident, problem, and change management. That means they actively participate in *on-call* schedules for critical services downtime as first responders. They need to isolate the faulty components of the service, troubleshoot the causes of the component issues, repair them or provide a workaround, reestablish the affected service to nominal performance, and verify whether the service has been restored from the user's perspective. After significant service disruptions, SREs must determine their root causes and contributing factors. They implement change requests to the systems, backing services, delivery pipelines, integrations, infrastructure, and applications.

Second, they are accountable for maintaining systems, services, applications, and infrastructure. They may need to patch a bug into production or assist the development team. SREs may have to deploy a new software version using a **canary release**, **A/B testing**, or **blue-green deployment**.

Third, SREs have the responsibility of taking care of the observability platform. That includes installing, configuring, maintaining, and monitoring the observability tools. Yes, we monitor the monitoring.

Engineering work responsibilities

SREs do engineering work to reach higher levels of availability, resiliency, performance, quality, and scalability on a system. They work on each configuration item or component to increase its reliability. The overall system delivers more trustable services and SLOs by handling each component reliability index.

Site reliability engineers are responsible for reliability metrics, such as the **mean time to detect (MTTD)** and **mean time to repair (MTTR)**. MTTD indicates how fast the monitoring system can detect a service problem or an anomaly that will lead to a problem if nothing is done. MTTR indicates how swiftly an incident is repaired after it's detected. Those metrics make SREs accountable for the effectiveness of the observability platform and tools, and the runbooks documentation.

The **mean time between failure (MTBF)** is another reliability metric under SRE accountability. That indicates how much time it takes for a system failure. SREs must adopt the blameless postmortems principle to improve this metric every time a failure happens. And that translates to multiple reliability enhancements to different parts of the system as a result of these postmortems.

SREs are accountable for toil management. The less toil we have in systems management, the better the metrics mentioned previously. Site reliability engineers work tirelessly to detect and eliminate repetitive tasks devoid of business value.

We described the ordinary responsibilities of an SRE with the intent of giving you an idea of what to expect in this career. Of course, this is not a comprehensive list of duties or intended to suggest a constraint to their responsibilities. As long as they work to fulfill the guiding principles, they are doing SRE work. We are going to review which activities SREs execute daily next.

An overview of the daily activities of an SRE

Now that we have examined SRE responsibilities, it's time to check what you, as an SRE, should be performing on a frequent basis. There's no better way to understand a profession than by asking what someone does in it. When you go to a job interview, you probably want to know the activities a person in that position will carry out. SREs will have a list of assignments as sticky notes on their displays. We have separated those notable activities into two sections:

- Reactive work activities
- Proactive work activities

We'll start by understanding reactive activities.

Reactive work activities

SREs execute many tasks that don't lift (or shift) system reliability directly; they are usually operational types of work. Nevertheless, those activities either lessen the service downtime or mitigate risks. Examples of jobs that SREs perform daily in this category are as follows:

- Repair or restore a system or multiple services to their original state
- Follow and execute instructions from a *runbook* (standard operating procedure) during an incident to diagnose the application
- Implement a change request to apply a patch to a software component

- Attend a meeting to run a postmortem with system administrators and developers about the recent service or system outage

- Install a new Kubernetes cluster for a new application according to the development team's specifications and enable monitoring of it

- Configure a new cloud-based service for a new application following the architecture design and include it in cloud monitoring

- Deploy a new software release to VMs and execute the testing scripts

Proactive work activities

SREs also carry out jobs that improve the quality, scalability, observability, manageability, resiliency, or availability of a system or service. Since those tasks increase the reliability levels of specific systems or services, they are considered proactive and mostly engineering type of work. Such assignments affect *toil* and technical debt. Examples of this category are as follows:

- Maintain a *runbook* on how to diagnose problems with a specific application

- Design and develop an automaton to execute procedures previously documented in a *runbook* automatically

- Establish, together with the DevOps team, the release strategy, such as a **canary release**, **A/B testing**, or **blue-green deployment**

- Work with the SWE to add management code to the application so SREs can instruct the application to do self-administration or self-healing operations

- Work with the development team to adopt an immutable infrastructure philosophy into the application-building process

- Instrument the application code to increase its observability with logs and traces

- Design and implement observability to obtain good metrics, events, logs, and traces from a critical application

> **Note**
>
> Site reliability engineers perform many more activities than the ones listed here. This is not a comprehensive list; the only intention is to show you how SREs work across multiple dimensions and aspects of systems and services.

We listed what an SRE does frequently. We wanted to give you a good sense of their day-to-day activities and how it differs from other roles. Again, this is not a complete or closed list. We want to close this chapter by telling you who our SRE rockstars are.

People that inspire

We want to finalize this chapter by pointing out other SREs that have inspired us and have been encouraging the wider community. We couldn't even think about starting this book without the work of the parents of site reliability engineering at Google. We are immensely grateful to them. Site reliability engineering would probably not exist outside Google if they had chosen not to share their thoughts, principles, techniques, and practices through the site reliability engineering foundation books. They are mandatory reading for anyone following this career path. If you haven't read them yet, please check out Google's site reliability engineering books at this site: `https://sre.google/books/`.

We want to recognize a few other rockstar SREs that have really made a difference in our professional lives as individuals. They are trailblazers of site reliability engineering outside Google.

Jeremy's recognition – Paul Tyma, former CTO, LendingTree

In technology, finding your way can be difficult. The constant struggle of being an SRE leads us into discussions of what went wrong; often, we have to say what some don't want to hear – that a negative thing happened due to what a person or team did or didn't do. We are, in fact, often the bearers of bad news. Paul opened the door for me to become an SRE, and we drove a great reliability revolution together. Most importantly, he taught me that there is a balance to all things, and we have a choice in that balance. And what we often consider a responsibility or duty can have its limits.

Rod's recognition – Ingo Averdunk, Distinguished Engineer, IBM, and Gene Brown, Distinguished Engineer, Kyndryl

Ingo and Gene triggered a small revolution inside IBM by designing and deploying site reliability engineering principles, practices, professions, and methodologies to its organizations across the globe. They first transformed many internal teams to adopt such extraordinary tenets, then later, they helped external customers in doing the same. Of course, they didn't accomplish this alone, but they were (and are) paramount examples of technical executive leadership. They shaped the site reliability engineering profession from within IBM, which later spread to Kyndryl after its spin-off.

Summary

In this chapter, we learned what the site reliability engineering persona looks like – what they know and how they think. We divided their mindset into smaller traits so you can understand what's expected of you on this journey. Also, we explained why the site reliability engineering profession is unique. Then, we entered the practical knowledge sections and looked at their main responsibilities and daily activities.

By now, you should be able to explain why someone would become an SRE and the path for that. You can differentiate the site reliability engineering job role from others and understand what the site reliability engineering persona is. Finally, you know the typical duties of SREs, and the activities most SREs perform. As we close this chapter, we hope everything we say here resonates with your career aspirations.

In the next chapter, we dive into the world of numbers and understand how they are intrinsically part of the site reliability engineering domain. You will learn why SREs see production systems like no one else.

Further reading

You can read more about the site reliability engineering profession and other topics in the *Awesome Site Reliability Engineering* repository: `https://github.com/dastergon/awesome-sre`.

2
Fundamental Numbers – Reliability Statistics

Numbers are all around us, enticing us to buy and trust, and even affecting how we define success. They give us a sense of something quantifiable and invoke emotion and spark revolutions. As a **site reliability engineer** (SRE), numbers tell stories and can give us a sense of joy. Yet measurement isn't always as easy as putting a ruler against an object in today's world.

In this chapter, we'll explore the value of reliability statistics and how sometimes perception is just as valued as the raw numbers themselves. We'll examine the true value of a **service-level agreement** (**SLA**) and how it's often not as you would expect. We'll explore what that means for external partners, and why inexpensive services going down can cause immense revenue impacts. We'll give some words of advice on how and when you can negotiate when SLAs are not met. In fact, the SLA and the business around its creation are vital to how these metrics are addressed, so we'll spend a few pages discussing SLAs specifically.

A measurement or indicator referred to as the **service-level indicator** (**SLI**) is paired with the **service-level objective** (**SLO**), which, together, define health. For example, if we use the error count per hour as our SLI, our SLO may define that over 20 errors per hour, we get an email alert, but over 100 errors per hour, we immediately send a page to an engineer on duty.

Another way to think of it is that an SLI is a measurement, like a child's temperature. The SLO is how we use the SLI to define health; for example, a temperature over 99 degrees is unhealthy, but a temperature over 103 is an emergency. Finally, the SLA is an agreement between parents on the actions to take at what temperatures, and with a doctor and hospital on what defines an emergency response.

Additionally, we will define and discuss the **mean time between failure** (**MTBR**), which defines how often failure happens. And to define the total time the organization takes to resolve outages, we dive into the **mean time to resolve** (**MTTR**), which is a metric of the time required to resolve issues.

Finally, we'll explore the human side of statistics and how numbers can both drive change and, when levered improperly, even grind progress to a halt. But perception can sometimes be more valuable that the metrics.

In this chapter, we'll be exploring the following topics:

- SLA commitment – a conversation, not a number
- Defining and leveraging SLOs and SLIs
- Tracking outage frequency with MTBF
- Measuring the downtime with MTTR
- Understanding the revenue and customer impact

SLA commitment – a conversation, not a number

An SLA is often advertised as a single percentage, such as 99.9% uptime. This singular view of an uptime metric for providing service is a simplified view that is riddled with exceptions.

Internal partner SLAs

SLAs should be viewed as the commitment to service that starts at a business level and discusses a technical process. By bringing the business into this discussion, we learn the importance of the processes our technical solutions implement. Those discussions often include multiple partners, each with their own agenda and technology and business needs. The preference is to always include SLA discussions as part of the initial planning of new services or features, but it's often overlooked, or worse, assumed that the service will always be available.

Starting these discussions opens a discovery into individual services, assets, and external partnerships. For example, a database often stands apart from the services it works with, or different microservices may have their own needs to discuss. Together, the group explores a number of different vectors and common needs, identifying the needs the company wants to fulfill, or what experience and revenue expectations we have by offering custom-made hats to customers.

The questions I start with are fairly simple:

- What is the impact on the business if this service fails?
- What is the customer experience if this service fails?
- What can we do to improve the impact or experience?
- What other services and providers do we depend on?
- Who depends on our service?

Let's look at how that conversation could be different for two different areas of a website that makes custom baseball caps. The company's website takes live orders and uses a microservice to calculate the manufacturing time and cost per hat, a process required to generate pricing for the customer's potential order. A nightly process creates shipping labels and emails tracking numbers to customers.

During the discussion about the manufacturing analysis microservice, the business identified that if the service goes down, a customer would not receive their pricing, and would likely leave the website. This would be both a negative experience for the customer and a negative impact on revenue and the brand. This microservice clearly needs to be a highly reliable process.

We compare this to the software run nightly, which creates shipping labels and emails tracking numbers to customers. Business impact is almost non-existent if this processing is delayed a few hours. Customers naturally want tracking information quickly in these times of e-commerce; however, a few hours or even a day's delay is not very impactful, especially when making custom hats. This process is quite different from one that must be highly available.

When we look at the SLA, the commitment to service for these two processes is extremely different. One must include highly available architecture and support **zero downtime** (**ZDT**) deployments, and the response time for any issues that arrive needs to be extremely fast. This SLA may also include a requirement for the application to be written with a higher level of effort, including more logging to help identify issues faster, better exception and error handling, and multiple retries. In comparison, the shipping label creation process response time would easily allow an engineer to grab a cup of coffee and have time to dig into logs and processes. The SLA would note a longer resolution time, and perhaps even alternative paths, such as making labels manually. Being able to keep the warehouse staff productive and customers notified may be the most important part of this SLA.

Either case requires us to look at the business process, and this is the heart of any good SLA discussion – bridging the gaps between business, technical, and cost considerations.

External partner SLAs

External partners bring immense value to an organization and provide services ranging from internet connectivity and cloud infrastructure to data processing and reporting. Even business functions such as HR and warehouse tools are often purchased rather than built internally. Some services may be governed by the **terms of service** (**TOS**), which are often very rigid and provide the same service agreement to all its customers. More complex services are often negotiated between companies, using a **master service agreement** (**MSA**). When reviewing these documents, I highly suggest consulting your corporate legal department.

TOS

Found primarily in services with a large customer base, the TOS defines what is provided to all. Designed primarily for consumers and not businesses, the TOS often goes out of its way to provide protection to the service provider. In fact, read through one or two and you'll quickly understand that often, you're paying for something that has absolutely no guarantee. And while customers would leave a provider who doesn't provide over time when your internet is down for days, the TOS often provides a consumer absolutely no recourse, no credits, and no promise of any type, and calling customer services becomes more about providing a courtesy credit as part of doing business and keeping phone call time low.

One part of the TOS that has recently been changing is in terms of consumer privacy. In light of the EU's **General Data Protection Regulation (GDPR)** and the **California Consumer Privacy Act (CCPA)**, new regulations in privacy have forced companies to change the way they interact with information about a consumer. Beyond being responsible for asking whether we will allow cookies on a website, these regulations provide better protection for consumer information – and this is often clearly addressed in the TOS.

MSA

In comparison, an MSA can be exhaustive and often includes lawyers from both entities. Resembling more of a contract negotiation, the language can get exacting and both sides often provide requested changes. These documents not only act to protect companies but often contain restrictions regarding customer data, interactions with employees, and even open-source provisions.

The cost of more 9s in an SLA

We can't talk about SLAs without talking about the 9s. When SLAs are quoted, we often hear it as a number of 9s, for example, 99.999% uptime (that's 5 minutes of downtime per year) – an impressive number that most in leadership would love to hit.

I've sat in a fair number of meetings where we state simply that the reoccurring multi-million-dollar cost of implementing and maintaining a 99.5% to 99.999% uptime has no cost benefit when the downtime costs are in the low 6 figures per year.

A final word on SLAs

We've explored many aspects of the business of SLAs, and that's the important part to remember. An SLA is a promise of service between technology and the business. It can be as simple as an uptime or an acceptable percentage of successful versus error transactions or even defining response times, level of effort, and expectations in the development of software and observability tooling. The most important part of an SLA is the discussion and understanding of both the need and cost to provide it.

Next, we'll discuss how to define measurable goals in the form of SLOs and SLIs.

Defining and leveraging SLOs and SLIs

SLOs and SLIs define how we measure health (the SLI) and what is considered healthy (the SLO). A very popular SLI is uptime, a measure of the amount of time a service or system is able to do its job as requested. I often say that the SLI is like a person's temperature, and the SLO is the range of temperatures defining healthy – for example, 106 degrees is my SLI, and knowing that anything above 104 is an emergency is the SLO.

Now, we often see an SLO identified as where we want to be, the healthy zone we strive for – but not defining levels of health as part of the SLO just makes sense. After all, there is a very different response required for a 99.5-degree temperature and a 106-degree temperature, because the levels of health are often not singular.

I refer to SLI as the *measure of health* because it defines a specific value that can be directly tied to health. SLIs should be very specific and, most importantly, measurable. When defining your SLIs, attention should be paid to how the system generates this data, and possible outside influences that could skew the results.

One example of commonly used SLIs is HTTP result status codes that provide a common way to indicate success or failure when interacting with a web server. This SLI is often also used to define uptime. Another popular SLI is error rate, or how many errors happen in a specific time period as a count or percentage of total transactions.

SLI skew

SLIs can often be impacted by noise from the real world – those impactful events that are often not seen while developing or even thinking about the perfect ecosystem of a system.

My favorite skew topic is web server (HTTP) result codes, especially when used for uptime calculations. Status codes provide a quick and simple way to provide real-time feedback on what's happening, especially to build alerting. Where we run into issues is when you don't allow for some variance, and in the world of uptime, every tenth of a percentage can impact the perception of the reliability of your website.

First, and most impactful, is bot traffic. Search providers have farms of bots scouring the internet, going from site to site, to both absorb web content for searchability and gather links to enable crawling more sites. With this strategy, we begin to identify a few interesting ways in which error-type result codes can be produced:

- By following links in the search provider's database that may have changed or have URL parameter options that may be invalid and cause errors
- By looking for standard files on web servers, such as `favicon.ico` or `robots.txt`, which you may not have created
- Health checks as part of observability, load balancers, or container management
- Internal functional testing that causes error states intentionally

How much can this skew an SLI? In my experience, as much as 2% – which is the difference between 98% and 100% uptime. That small 2% is greater than the allowable margin of downtime of most commercial systems.

Status codes are not the only areas we can see skew; error rates exhibit their own unique skew. One of the biggest struggles is errors that are resolved by retries – if an error happens, but 15 seconds later, a second attempt works, is there any impact? Even worse, tracking these unique events can be hard to track and requires unique error capture and logging in the code. We would want to log if an error causes a retry, as we should track these; if a retry is successful, it's not often considered an error.

Beyond retries, the definition of *error* can be subjective. Take, for example, a process that retrieves information on a co-signer of a loan. When the process fails, do we call that an error? Well, let's look at the process. Not all loans have co-signers, so is this an error if there is no co-signer on the loan? Of course not. I have seen a number of these types of issues when working with errors – and not all errors are actionable or very impactful. Choosing error levels and understanding individual errors can have a large impact on error rates in general.

Finally, an SLI needs to be aware of scheduled maintenance windows, where business and technology partners agree on impacting business to perform actions. Is this part of the SLI calculations? And more importantly, can the system used to track your SLIs make the proper adjustments to not take planned downtime into account in calculations?

Examples of SLIs

In our previous example, the website creating custom hats for customers, we can break down each of the two processes – a microservice used in pricing orders, which requires high availability, and a nightly batch process generating shipping labels and emailing tracking information to customers, which, while equally important, had a wider variance in its uptime requirements.

An SLI we might explore for the microservice, at first, you may think is as simple as the error rate in association with the presentation of a price to the consumer. This would directly illustrate the microservice is working, but would typically include other parts of architecture, including the automation in the frontend that retrieves, calculates, and displays the pricing. Looking closer at the microservice, we can look at the HTTP response codes on the API, but the skew of possible external traffic can cause another skew. We could use logs from the microservice and log successes and failures, but if the API has issues, the errors in that layer of the architecture won't be visible in the SLI.

As you can see, finding a proper SLI for the application that gives you both a complete view of the process in question while having the least skew is difficult. When I look at this challenge, I would capture the errors in the logs against the successes and pair this with the success rate of generating pricing.

Compare this to a backend service to generate shipping labels and email tracking numbers, which can withstand a few labels failing to be created, where the labels are printed manually by warehouse staff. And while we want to keep manual label creation to a minimum, a few a week is likely very acceptable. In this process, the count of unsuccessful labels created may be a worthy SLI.

SLOs

SLOs are the bridge between our SLIs and SLAs, and they define health. But SLOs are not just about the agreement: we actively use our SLOs during day-to-day validation of system health and stability in the form of dashboards.

It's not just day-to-day activities; SLOs play a major part in the diagnostics and validation of system outages. Beyond knowing what part of a system is working as expected, understanding the relationship and trending of SLOs gives a sense of the state of the outage. When we look at the raw number driving an SLO, often displayed as an average over time, we lack motion, making it impossible to see whether conditions are getting worse or better. This is one reason dashboards are so beloved: a proper dashboard can not only show health from a numeric standpoint but also show the motion defining how an outage may be evolving.

SLOs and time

One important aspect of SLOs is the element of time. Longer time spans give us a sense of stability but lack the immediacy you'd want for alerting. On the other hand, presenting SLOs in a report may require looking at statistics over time.

Why is time so important to SLOs? Consider a goal of 99.9% uptime; when we look at a 24-hour day, that allows for 86.4 seconds of downtime per day. But when we look across a month, it balloons to 43.2 minutes, or 8 hours and 38 minutes per year. As time stretches out, outages can be distributed across months or a year and made to seem less impactful. And while it may give a perception of low impact, let's look at the revenue loss on a website that generates $2,000 in sales a minute. A single outage in a year gives you an uptime of 99.94% – a rather impressive uptime. But when we look at the revenue loss, it's $600,000. Quite a difference in perception – 99.94% uptime, but still a half-a-million-dollar loss.

As we consider SLOs and SLIs, we should consider the revenue impact and time spans. Each plays a unique place in the representation of the state of systems. Most importantly, ensure business and technology understand and agree on how these metrics impact both revenue and the customer.

Utilizing SLOs and SLIs in constant improvement

Beyond just knowing the state of our systems, improving their stability is often considered the basic job of an SRE. Understanding how these SLIs can and do change over time is important, especially as we add new layers of observability. Just like a closed box, the act of viewing its contents always impacts the integrity of the box. SLIs can be skewed by the addition of observability.

When we look into the improvement of systems in general, it's tempting at times to update the SLI models to remove more skew, and thereby, artificially inflate our measurements to meet our SLOs. When we talk of constant improvement, it's important we call out these measurement differences, and while an improvement in measurement is a worthy find, it must be set apart from the improvement based on higher reliability.

As you make the journey to start improving, you will often find that it's two steps forward and one step back; an improvement in one system may highlight issues in another. In fact, when building observability in a new system, you should expect to find issues that may sometimes have existed since the system's inception. Indeed, defining improvement is often not as simple as turning a dial – it's a slow and sometimes painful process, but understanding how to talk about your wins in terms of your SLI and SLO gives you the power to bring to light the small wins in battle.

So we can now define the health that we use to measure our system's capabilities, but reliability is about failure – and just like a car, the more it's in the shop, the less we enjoy it. The MTBF and MTTR define both how often we have to put our car in the shop and how long it normally stays there.

Tracking outage frequency with the MTBF

While SLOs and SLIs give us an understanding of the health of our systems, the **MTBF** illustrates our system's frequency of failure. Understanding how often failures happen not only helps to be more prepared in terms of staffing resources but also stabilizes the revenue and trust of a brand. These high failure rates can also impact leadership's confidence in engineering.

To calculate the MTBF, you must first choose a time period to look back upon – for this example, we'll use 12 months. Then, we count the number of failures during that time, so 6 failures divided by 12 months makes your MTBF 2 months between failures. It's important inside an organization for all teams to use the same time period to ensure the comparison between departments is accurate. Even further care needs to be taken in the definition of what a failure is: is it a complete system outage? Just a partial outage? Or just a database connectivity issue for 2 minutes? Defining these standards will help ensure the MTBF of one system is equally comparable to another.

As a fundamental number that measures how often a system or application goes down, you'll be interested to know that there are two different ways the industry has chosen to define a "failure," and thus two different ways to look at the MTBF. The first is by defining failure as times when we must put our hands on keyboards to resolve issues. The second is by looking at the time between the errors a system generates. Both are very valid measurements of reliability.

Next, we'll venture into measuring the length of time we are in a failure state and break that time up into the essential parts of the failure response.

Measuring the downtime with the MTTR

The **MTTR** is the average amount of time an issue takes to be resolved. It is generated from the average of that time span. The MTTR is often thought of as response time – how effectively a fire department can get to your fire and put it out. It is also the amount of time we are often impacted by each outage, so when the MTTR goes up, we often see a decrease in revenue and customer satisfaction. The MTTR has multiple smaller elements inside of it, each contributing to the overall outage time. Let's step through a typical outage and quickly examine each of these elements:

- **Detection time**: The time between the outage start and when someone noticed it. This often starts with the root cause and measures up until the first person or automated notification says that something is wrong.

- **Notification time**: The time it takes between detection and when engineering assets first respond. This could be the time it takes for someone to acknowledge via phone or a messaging system, a page being acknowledged by an engineer, or even days later, when an email is finally read.

- **Response time**: The time between knowing an issue has started and engineering responding to the issue.

- **Diagnostic time**: The time to find the cause of an issue. This resolution may be an actual fix to the root cause, or a band-aid put in place to temporarily resolve the issue.

- **Deployment time**: The time to perform a deployment.

- **Validation time**: The time to validate the resolution addresses the impact. It should be noted that I said *impact* and not *issue* – it is important to call this out because our goal is to resolve the impact on the revenue and customer, and addressing the actual issue may take days or even months to properly resolve.

Each of these steps introduces challenges to the total outage time and there are systems and processes you can use for each of these to decrease these times and resolve issues quicker. Outage management itself is so important that we dedicate an entire chapter of this book to it.

I should mention one other common question with the MTTR: when do we actually start and stop the clock? With technology today, we can often pinpoint the exact moment an outage starts – for example, the moment a hard drive ran out of space, or a database or system went offline. However, the technology of today is far more complex than a single server, especially in a highly available, multi-server environment – one entire server being down can sometimes be tolerated for hours before the traffic starts increasing, and increasing load overwhelms the servers still running. In this example, did the outage start when the server went offline or did it start when the revenue and customer experience started to be impacted?

Companies are in business to generate revenue, and while we talk about how to measure outages in time and frequency, it all boils down to how it impacts both the customer and the revenue – something we'll dive into next.

Understanding the customer and revenue impact

More than most, I talk about SRE in terms of customer and revenue impact. In a world so heavily dependent upon the services that prop up websites and technology, it's not difficult to understand that when they go down, we feel the impact in everything, from watching our favorite show to airline flight cancellations, and even something as simple as purchasing gas. Even worse, when customers feel abandoned or under-served, they often look in the marketplace for alternatives, and this tendency increases the longer and more often a system goes down.

Beyond the customer impact, companies often feel financial impacts on their bottom-line during outages. This impact ranges from simple lost sales to more complex erosion of a brand, which can cost advertising dollars to fix and stifle the growth of a company. We also often forget the human element of outages; having been on calls that cost companies thousands of dollars an hour in manpower, it's easy to see how having a dozen people responding to an outage for hours can quickly impact the work of the day, pushing deadlines out and emotionally straining staff. Even worse, when these responses happen during the off-peak hours, we can impact an employee's family time, stealing that precious time often used for an employee to recharge. I've even witnessed first-hand staff leaving positions, simply due to the undue burden of after-hours response, which, in turn, can cost tens of thousands of dollars in the recruitment and training of new engineers.

Transparency in outages

As we've discussed, the impact on a brand, customers, and business can be extremely complex and wide-reaching, but time and again, the market has proven that the single largest tool we have is transparency. The obvious one is transparency with management, with ensuring that we as engineers give accurate timelines for the resolution being the most important. And this can be challenging, especially when you have a group of your top talent digging into search engines for more than a few minutes or heavily engrossed in code attempting to resolve an issue; sometimes that timeline is not a simple task.

Transparency with our customers as well is a must, especially for customers with long relationships with a brand, sometimes passed down through generations, as they are much more willing to stay the course when they feel valued enough to be given the truth. Often cited as the use case in customer transparency is the Tylenol recalls in the early 1980s – a Johnson and Johnson brand easily recognized around the world. One bad actor in the Chicago area replaced Tylenol capsules with poison, killing seven people. Tylenol's response to the incident has been a model for customer transparency. Immediately, the company performed a $100 million recall of all Tylenol products on the market and reinvented overnight the safety sealing of their products with multiple layers of protection from tampering.

Tylenol, often cited as the eighth victim of the killer, never swayed from its responsibility to remove potential harm from its customers – in fact, driving consumers to demand taper protection as a minimum requirement for consumer food and drug products across multiple industries. In short, Tylenol was honest, truthful, and more importantly, took immediate action out of a sincere concern for its customers. Finally, Tylenol on its own initiative went a step further, offering to replace any existing bottle of Tylenol a customer may have in their medicine cabinet, and vowing to destroy any potential harm.

We often cite drops in sales numbers or the number of customers who we are not able to get online service due to an outage. But the damage to a brand has a far-reaching impact you can often not put a value on, stifling the growth or, even worse, reducing brand loyalty due to lack of trust.

The rockstar SRE's SLA

As a rockstar SRE, when I discuss SLAs, it's not only to talk about uptime – in fact, the uptime is rarely discussed. I often start with a discussion on SLIs and SLOs, bring in response expectations and time frames, and finish with communication.

The first question is, what is measurable that would define health from both a business and technical perspective? Those are our SLIs. Bringing in business partners is imperative because a single error in a nightly import routine can be extremely impactful, while some services can withstand hundreds of errors a minute without impact. That definition of what is healthy (our SLOs) should drive the type of responses. Do we give our child Tylenol for a high fever or rush them to the hospital?

Starting with the basics of who responds and in what timeframe, we identify the level of effort as defined by a measure of health. Together, we define how soon the response is expected and in what timeframe – and even an escalation policy for when staff members are not available. An example of this may be the network operations center is expected to engage within 5 minutes of an alarm, but if they do have an issue resolved in 20 minutes, senior engineers are paged to assist with the response.

We have these discussions and set response expectations for a number of reasons, but the most important one is setting the correct expectations. This can be as simple as staff understanding what their responsibilities are, when and how quickly they are to respond, or even when they should reach out to additional resources.

In short, we want to define a path that is acceptable to all parties, and it's okay to come back together to discuss a needed change. In fact, I encourage a touch base with teams you support on a quarterly or every 6-month basis.

Summary

In this chapter, we explored the importance of an SLA and its creation, which, as we remember, is more about the relationship between business and technology. Even though we don't always have a say in our SLAs (especially for TOS, base offerings, or existing MSAs), conversations with your vendors focusing on what is best for both companies to be profitable often lead to more trust and even provide paths that not all customers may have to resolve issues. We also explored how not all SLAs require immediate action or 100% uptime, and how potential impact should be the driving force in the response.

As we dove into the SLI, the *measure of health*, we reviewed not only how finding the right way to measure is important but also how outside influences can impact getting a correct measurement. We also took a look at SLOs, often used as a tipping point for working versus broken, and how their view with respect to time is so important.

And remember, an SLI is as simple as an indicator of health like a temperature. The SLO is how we define health, for example, the temperature at which we agree a child is unhealthy, and the SLA is an agreement between parties such as parents or doctors, defining what action is taken and when.

Finally, we spoke about outage statistics; MTBF and MTTR play a significant role in the perception of effective outage management, breaking down the MTTR into parts so we can target the ever-present journey of constant improvement. Don't worry, in *Chapter 14, Rapid Response – Outage Management Techniques*, we'll spend a whole chapter on outage response.

In our next chapter, we'll discover the business of building software, and how this can result in poor reliability and lack of observability.

3

Imperfect Habits – Duct Tape Architecture and Spaghetti Code

So why is it that when we look at repository after repository of code written today, we see so many examples of poor resiliency and, honestly, poor coding practices? Why are retries just honestly forgotten in most code written? And finally, why are concepts such as writing a good logging statement just plain missed in so much code?

Then there is architecture. How often has a simple HTTPS certificate expiration taken down companies for a day or two, or even weeks? And why do so many think their systems are 100% resilient, especially in the world of cloud computing and the newest fads of serverless?

When you've seen the amount of code and architecture I have in my career, you start looking beyond the code and start looking at why the industry allows such low-quality code to exist; in fact, in some ways, asks for it! It's not an easy question to answer, and like most things in life, it's far more complex than you might think.

In this chapter, we'll dive into many of the reasons poor-quality code exists, from businesses to coders, to colleges and beyond. We'll also look at ways to combat poor design, and poor coders, and even take a few moments to examine why it's so hard to fire those who often produce little and know even less.

By the end of this chapter, you will be able to understand the business process of building software and development. You will also understand the reasoning behind 100% uptime and how to move the needle in multiple ways in an organization to combat poor code and architecture.

In this chapter, we will cover the following main topics:

- The business of software development – let's start with the dollars
- The A/B testing mindset – the art of change in customer interaction

- Dedication to the craft of development – and why some are just here for a job
- Reviewing the merge request – it's about training, oversight, and reliability
- Why business wants us to outright ignore best practices
- Mixing good and bad – tricks to wrapping bad code and making it resilient

The business of software development – let's start with the dollars

The basics of business are simple. First, make sure your revenue is more than the cost of the product and service you provide. For simple transactions, such as a person buying cheap tools from China and selling them at double their cost at a swap meet, this is cut and dry. But when you're talking about a multi-billion-dollar company, the differences in how value and revenue are calculated take on a vastly different mathematical equation, which often ends in a benefit of savings for a company. We'll talk about different mechanisms used, including a few that might shock you – and even talk about how capitalization of software is used to drive the cost down – well, kind of.

Defining the "value" of software to a business

So many times, people ask, "*Is this worth it?*" – from travel excursions to new keyboards, a fancy new kitchen knife, or even brand loyalty, the terms *value* and *worth* are some of the most subjective concepts in our lives.

Value is a strange thing and often has emotions attached. Take, for example, the brand loyalty of a family who always purchases a vehicle from a specific manufacturer. Value is not only in the brand, but it's influenced by both the trust inherited from their family and a sense of not wanting to disappoint a family member, be different, or worse, be teased. And what about those $4 boxes of cookies, $12 cups, and that catalog of overpriced dollar-store items all sold to us by the children in our neighborhoods? Is there more value in a $4 box of Thin Mints compared to a much larger $3 pack of OREO Double Stuff Cookies? When we think about helping out our local Girl Scouts, why does value increase?

In business, value can be impacted not only by the spreadsheets used to compare the cost to benefit but also by the beliefs in market forecasts, power struggles, and even ego. It's not unusual to see a project being built because of someone's ability to sweet talk the boss (or CEO), and likewise, software that has the highest potential value can be swept under the carpet.

The value of protecting business

So, let's talk about some examples of value in software. During the COVID pandemic, many consumers simply felt unsafe entering stores or other public areas. In response to this, we saw many companies offering more online ordering options and curbside pickup of products. And companies that didn't offer these new features suffered. Beyond training staff to provide these services, or partnering with companies who mass recruited part-time delivery drivers, software to drive these new processes had to be built. Yes, the software was built partly because, without these new features, companies increased their risk of failure, but as we came out of the pandemic – another value emerged – brand loyalty. Brand loyalty means two things to a business, less of a requirement to sell items at a high discount to get people in the door and more overall business. And it worked – how many of us have a favorite new restaurant or grocery store simply because of the ease of use of their apps, the trust we have in their ability to pick out vegetables for online grocery orders, or just because we discovered a new great place to eat that we love?

In this example, the value is in keeping the business open and protecting revenue. But for the companies who executed this flawlessly – they stood to gain much more in brand loyalty, which drives growth and protects revenue streams.

The value of growing a business

We've looked at the value that protects a business, now, let's look at the value that is designed specifically to grow a business. Let's take our favorite online custom hat company from *Chapter 2*, *Fundamental Numbers – Reliability Statistics*, and give them the ability for users to upload their own pictures to put on hats. Instead of just adding letters on the screen or having a selected group of pre-approved symbols, now the user can upload company logos for the next company picnic or even a favorite child's drawing for a grandparent to proudly wear. This value is determined by the cost of both building out the new features on the website and added customer service cost when an image won't upload against the possible increase in sales. Is it a bet? Yes, in fact, that is the real definition of innovation, betting on what can work and knowing how much you can afford to bet.

The value of saving labor costs

The third type of value we'll explore is cost savings – and it's getting more and more popular! Ever ordered from the McDonald's mobile app? Or had your question answered in one of those online chats on a website? Believe it or not, both are designed to allow a company to make more money by reducing your interaction with one-on-one staff and increasing customer satisfaction.

Imagine the time you can save in a drive-through line if you don't have to both look at a menu and choose what you want for lunch – or, worse, wait for your partner to choose! By moving that process to a mobile app, my partner can enter the order while we're driving to pick up food, thus reducing our wait time and ensuring that the pesky speaker doesn't get our order wrong. This flow also reduces the time an employee has to interact with customers, reducing labor costs. Another great example in the customer service industry is the chat option, which often even includes AI-driven bots answering your questions. When you do get to an actual agent in chat, they are using the time you take to answer their questions to help others, allowing multiple users to be serviced by one agent.

While labor cost reduction is a favorite way to save revenue, it's also not uncommon to see small changes being made and tested against current processes, to improve the revenue stream as well, this is A/B testing, and we'll discuss it next.

The A/B testing mindset – the art of change in customer interaction

A/B testing is simply tracking how customers respond to differences in the process. Most often employed on websites, we've also seen A/B testing showing up in process differences at different stores or even selling new products in *select markets* in fast food chains. The concept itself is simple, track the difference between revenue and customer experience outcomes based on differences in the process presented to customers.

A/B testing in customer flows

The most popular form of A/B testing is in the flow of customers on a website. Be it the collection of your personal information so a salesperson can call you to offer a product or services, the placement of the upsell accessories on a website when you buy a TV or toaster, and even the color schemes and graphics on a website.

The first part of A/B testing is the ability to track customers – and it's not as easy as it seems. For website-based customer flows, we often find customers visit multiple pages or even come back to the website days or even weeks later. So, websites leverage cookies, or settings, stored on a user's computer by the web browser to carry a unique customer identifier from page to page, even if you return weeks later. These cookies allow a company to track what pages you have looked at on a website. For example, the information on my cookies for websites would tell you I like looking at 3D printing and electronics, while my partner's cookies might say they enjoy looking at pet care products or Lego – it can also tell a company if you like using coupons or are the type of person who just pays full price. This history can not only help a company target specific users with ads but also define flows and even upsell experiences for customers on websites.

The second part of A/B testing is the test itself, the split, that sends some customers to version A of a website or process while others may see version B of the website. The changes could be as simple as version A has the old background and color, while version B includes motion graphics and a new, brighter color scheme. And A/B testing can take on many different variations, from color, to how a question is asked, or even completely different experiences and product offerings.

Analyzing the results of A/B testing

As you can guess, A/B testing is all about the results. Did I sell more if I used a new flashy website? Is blue or green the best color for buttons? Should I ask more or fewer questions when collecting information before I call you to sell you new gutters for your home?

The results are the key – what drove more business. Some companies are fanatical about A/B testing because, as you can imagine, a half-percent increase in sales can mean hundreds of thousands of dollars of increased sales in a month. Of course, the opposite is also true; a poor test result could also reduce revenue. But tracking the increase or decrease of the new test versus the original tells us whether the test works and defines changes the business should follow to increase revenue and customer satisfaction. There are even some in the industry who are willing to sacrifice just a little bit of customer satisfaction to increase sales.

Leveraging A/B testing to satisfy quarterly numbers

In today's business world, reporting results by quarters defines not only the worth of a company but can impact a multitude of opportunities and long-term value. In fact, missing just one-quarter of growth for a company can throw stock values into a dive and even impact ongoing negotiations with other companies and customers.

Since A/B testing is seen as a fast way to improve everything from direct sales numbers to lead generation and beyond, in short, it can be seen as a quick fix to improve revenue. And since we are working on a quarter-by-quarter basis, testing is often implemented quickly and can be revised often, driving change in rapid successions, sometimes even on a daily or weekly basis. Driving code changes this quickly can lead to short testing periods or even no testing at all. Pushing code out the door this fast can impact quality and leave little time for code reviews. Since a quarter is only three months, it's easy to see how time can be precious in this process.

Oddly, one result of a quarterly-driven numbers environment can be more mature in the area of production response. This makes sense, given the unrelenting push of code; production impact can be just as unrelenting. What's the saying? Practice makes perfect. In fact, some of the most advanced and capable production support teams can be found in these environments.

While reducing workers and performing A/B testing provides real, tangible results to the bottom line, we still often struggle with the basics of the employee, from a real understanding of why the level of effort is very different from person to person to how the job market negatively impacts companies. Next, we'll cover why understanding employees is essential to the reliability of the products they build.

Dedication to the craft of development – and why some are just here for a job

Engineers showing up for their job have a range of motivations. Understanding these is key to predicting the quality of engineering and enacting planning to build change within an organization. Make no mistake, the ability for a person to do their job well, and the quality of their work is not always coupled with a dedication to their craft or employer. This simple fact can be challenging in itself, and sometimes, the difference isn't always clear – to you or the engineers.

Today's technologists, including developers, are in very high demand. Between highly motivated recruiters, signing bonuses, and higher salaries, turnover for developers can be high. It also provides a unique management challenge – giving engineers freedom to do as they wish in many cases. As you can imagine, these engineers can require more than just a memo to bring them around in the world of better coding and reliability. Beyond simple issues of scarcity in the market, people can have social interaction issues, egos and communication skill gaps, which further adds to the complexity. And finally, some people simply believe their way is better – and belief can be one of the most difficult things to change in a person.

We also see two types of hiring practices; hiring a contractor allows a manager to make a single phone call to dismiss the engineer – often, we find this type of employee is often both less valued and less valuable, though that is not always the case. Secondly, we often see HR departments, especially at larger companies, work hard to keep employees due to the cost of using a recruiter or the pain of using internal recruiters and because they want to ensure all avenues are addressed before letting an employee go. Both of these issues, contractor hiring and HR stickiness, can make finding good employees and terminating poor-performing employees difficult.

In addition, often, new developers coming out of college can have huge gaps in their abilities, mostly, unfortunately, due to a good developer being able to make far more in the private sector – leaving most college professors to be less skilled in today's technology world. How bad is that gap? I've had college grads who never worked with cloud architecture, which is one of the skill sets I'd want an engineer to have at least an understanding of on day one.

So what does the relationship an engineer have with their job have to do with you being a rockstar SRE? Our ability to influence developers to improve coding habits and think in ways associated with highly available architecture, and even update a simple logging statement relies on our ability to sell the need to the engineer and understand what motivates them. And when we have to find other ways to implement new processes or policies, we need to be weary when we try to force change. Being a rockstar SRE means we know how to enact change in the least disruptive way.

A quick guide to communicating with your colleagues

As we've discussed, working with developers can be one of the most difficult things we do as SREs – and at times, we can feel like we've gotten nowhere – and, worse, that nobody is listening. The following list comprises some of the many approaches I've used to discuss issues with developers in my rockstar SRE career:

- Always greet developers with a fierce and hearty welcome! There is nothing better than a big smile and a handshake or wave over Zoom to make a person feel at ease.

- Get to meetings early and make small talk – especially in this Zoom-driven age of remote engineers – that short time at the start of a call to discuss the developer's child, dog, or new toy will go a long way in creating a bond.

- Take the developers out to lunch, and pay, or better yet, put it on your corporate card if you're allowed. Keep the conversation light, and always make it a point to ask them many questions about themselves.

- Approach problems in terms of gaps that you have a hard time understanding how to fill rather than describing issues as bad decisions. This opens the floor to discussion. On more than one occasion, developers have given me solutions I would never have thought of and were better than my own.

- When you have a production issue, be honest with the developer when you are struggling with their code or the process it handles – this vulnerability will allow them to own the code they wrote, and emphasizes their importance.

- Talk about process and code issues in terms of revenue loss and tie that back to the greater good of the company and our own bonuses – bring the issue into the context of impact and make it personal.

- Finally, whenever possible, let the developer be the hero, even if you lose the credit for fixing something.

And speaking of developers, an essential part of building strong, capable developers is done by providing insightful merge request reviews of their code, which we'll cover next.

Reviewing the merge request – it's about training, oversight, and reliability

Merge requests and code reviews are one of the most powerful ways to provide feedback to developers as they build code. Being one of the most valued ways to bring engineers to the next level, code reviews done by engineering leads and more than one, if possible, give engineers a chance to learn new techniques and technologies.

I have personally learned much from code reviews in my life, though, I thought the suggestion was more about style than content; it's beneficial to learn new ways of doing things and, most importantly, keep your fellow developers happy by taking their suggestions. As an SRE, we often do impromptu code reviews or have to review code during an outage – and we don't always like what we see, but it's important to keep an open mind for style differences.

Avoiding the typical rubber stamp mentality

Merge requests are a chance to review the work being brought into the next deployment environment and a time when we take a quick look at the code to see that it meets standards and looks to function properly. Unfortunately, many in the industry are simply rubber-stamping or approving merge requests without even looking at the code. This can not only be dangerous but robs our engineers of the chance to learn and grow. Most importantly, when we're moving fast, sometimes we miss something, like a prior commit being over-written by yours, reintroducing a fixed bug. I have had merge requests questioned because I missed something simple – and I'm thankful I work in an environment where code is reviewed before merging.

A word on production deployments

When we review **merge requests** (**MRs**) for incoming code or to deploy to other environments such as QA or production, we are empowered to say yes – and to say no. Remember, "*With great power comes responsibility,*" and MRs to deploy production code is no different. In this context, though, the bar is set much higher for what stops the merge. But stopping a merge request to production also carries with it far greater scrutiny by more than engineering, as often, product people are part of deployments.

So what are we looking for in a review? That's a very broad topic, but here are a few items I always look at:

- Does the function perform the task it appears to?
- Is there any case in which an endless loop could be caused?
- Is the logging reflective of what is happening and in the proper order? For example, does the log saying we retrieved data come before actually retrieving the data?
- Does testing truly test the functionality intended?
- Is there good error capture, such as try/catch, and equally sufficient logging of these errors?
- Is the code designed well in terms of reusability, or is the same function copied and pasted over and over again?
- Is the code easy to read?

During production deployments, it's always a good idea to review the MR prior to a release window – and when a problem is found, reach out to the developer directly instead of making a blanket statement to the deployment team when possible. And most importantly, remember, you are empowered to stop deployments – and I've said no before – but make sure your reasoning is sound and supported by others. I also always prefer to have another engineer on my team review an MR if I am going to reject it, especially in a big setting.

Why businesses want us to outright ignore best practices

We've spent a good many pages now discussing how value can be perceived and multiple examples inside a company of different types of value. We have explored A/B testing and how quarter-driven results can dramatically shorten A/B testing implementation time for developers introducing higher risk to code stability. And finally, we took some time to have code reviews and merge requests. So why did these subjects come up before we got to the heart of what this chapter is about? Great question – without understanding these subjects, we have no base to provide an understanding of what goes wrong in the development process.

The truth about the ownership of a developer's time

I made this mistake early on in my career – thinking that developers and their managers own their own schedules. I thought by speaking with the development team, I could ask them to assist with reliability and observability tooling inside their code and they were working things in when possible when they felt it was valued.

As a rockstar SRE, my tactics have certainly changed. You see, developers and their managers don't actually own their schedule or the tasks they work on – in most businesses, the product team owns the time of the developers. This discovery completely changed both my approach and ability to implement changes in code for reliability and observability. You see, when you talk with the product team, you talk not only about the little changes that need to be made but about the value the change can add. You see product teams work on simple concepts of what change brings value – and if it doesn't have value – they simply don't build it! And especially for large changes or shutting down a team's ability to deploy due to a high rate of deployment defects, honestly, if you haven't talked to the product team first, it's highly likely your plans will be shut down or at least met with a high level of contention.

And working with product teams can be impressive; not only have I seen product teams back SRE proposals, but I've had product teams on multiple occasions ask me and my teams to step into issues development teams are stuck on or help define solutions for projects. Not only is our team happy to assist in problem-solving in the development of projects, but we also welcome it – because, quite simply, it helps define how and what SRE can provide and strengthens the bonds with our product partners.

Most importantly, though, when working with large projects such as moving all your code and pipelines from GitHub to GitLab, your involvement with the project team and their understanding of the need for your actions coupled with understanding and discussing how your project can impact the delivery of a product is an absolute necessity.

Understanding the flaws in how we estimate development cost

In the software business, we often look at the time it will take to build a new feature or fix a bug. And this should only be done after we have a working idea of the process and solution so proper time can be allotted to the development. Then we provide an estimate to deliver a solution, though often we leave out additional wrap-around support, such as building new testing or interacting with DevOps to build a new database table or alter Terraform to deploy a new lambda function. Some of these *wrap-around services* can prevent deployment, such as needing a new database table. But others, such as additional QA testing build-outs, would not block the release of new software, and, if the resources are not available, it's easy to see how QA testing can be omitted from the deployment. The same is true for other items, such as documentation updates or load testing.

Software development cost should always include the wrap-around services, and this is best done by ensuring the definition of done, *a list or checklist of items required to call a task done*, clearly states items such as testing and documentation need to be completed before we call an item completed. A review of time should also include talking with service teams to ensure there is a clear view of both cost and build time requirements.

For larger projects where we may work with multiple teams inside and outside of development, coordination of development is extremely important when services rely on other to-be-built services. This planning, if not done correctly, can inflate schedules and even blow through deadlines.

So how do we protect the cost analysis of a project?

- Ensure the *definition of done* includes items such as documentation, testing, and observability.
- Bring in the partners for planning – and ensure those estimates are captured.
- When performing work after another team completes an item, ensure communication is kept open as to the state of that project. If it's slipping, ensure the project owner is aware.

Most importantly, when cost becomes a concern on a project, ensure that discussions about QA and observability are still happening – these are the golden tickets in ensuring smooth production operation. These are components that are easy to cut for cost or time reasons but severely impact the product.

There are hundreds of books about planning and estimating the cost of projects; the key for the rockstar SRE is to understand the basics of where estimates come from and the shortcuts and items that can be forgotten about, all of which can impact the quality and reliability of the end product.

Fast, good, cheap – pick one

This famous saying is true in almost every business. I've seen entire teams stuffed in a training room for weeks at a time, isolated to work on one major innovation – the managers even buying lunch to squeeze every moment possible out of the team. Or, worse, implement a weekend work policy. These types of arrangements, if not kept in check, can actually diminish worker performance not only from burnout; even the diet of pizza and fast food can cause workers to have less energy over time. When teams start to burn out or lose focus, we start introducing bugs and process analysis may go down, meaning you're writing to implement the process incorrectly. From an SRE perspective, we should be checking on these teams and even peaking at their code and commits to bring in observability earlier than normal into these types of projects to catch issues earlier in the development cycles.

When we look for cheap, it's not uncommon for inexpensive contracted development engagements to require additional oversight. And this isn't always due to poor performance. If one of my engineers completes a task but has questions that I do not have the bandwidth or process in place to answer, is the quality of the task a result of a poor engineer, or poor process and management? This is certainly something to think about with contracted development.

Since SREs are often tasked with production issues in third-party written systems, engaging early in the contract phase to ensure proper logging and observability is highly beneficial. Also, simply having the third-party team walk you through the process or flow can be a great indicator of the quality of the final product and allow early intervention if needed. You are likely going to support this item when it's delivered; as a rockstar SRE – we engage early and shape observability and examine risk along the way.

Why is observability the answer to reliability issues?

Something I've wondered about for years is that even when there are huge initiative drives to *build* dashboards and alerts, we forget to look at the root causes of reliability. Yes, observability is extremely important – even more so with poorly written code. But rockstar SREs should look at engaging themselves in multiple phases of the development cycles, especially to review logging and identify gaps in how we can view the workings of a process. We ask politely for little tweaks and updates along the way and offer newly tricked-out dashboards and assistance when possible. Finally, offering a dashboard for QA sites that becomes available during the first production deployments can be a lifesaver for teams – and allow your work to frame the team as the heroes when no deployment issues happen – and they can prove it with your dashboard.

The cost of highly available architecture

The cost to run highly available architecture can be overwhelming for some companies. Even worse, multiple cloud-provided services, for example, AWS **Simple Storage Solution** (**S3**), have no high-availability capabilities written in, so building one from scratch can be tricky and leave your system unavailable should a bug be introduced in that tooling. Indeed, running multiple servers for databases and even entire redundant data centers can be a very expensive endeavor. And yes, sometimes, the cost outweighs the benefit – so a company prefers to risk the downtime instead of spending the money for resiliency.

This is why the discussion about SLAs with your business is so important. Let's look at how this discussion about redundant architecture might look with our favorite custom hat website. You remember our two use cases, the microservice involved in pricing hat orders, which needs to be highly available, and the nightly label printing and tracking number email process, which can easily recover from multiple hours of downtime.

For the pricing microservice, we identify that not only does the service have to be highly available, but we also need the data to be available from multiple redundant sources. While the data must continue flowing, we make a cost decision that we are okay with the site not switching over to the secondary servers immediately, accepting a five-minute downtime window to allow servers to be built and come online in the second data center. This business compromise may save thousands a month in server costs by letting the secondary server lay dormant when not in use. In a startup environment; however, the time or talent may not be available to set up multiple data centers and automated failovers. The time requirements push reliability out of the picture entirely – another example of how business can impact reliability.

The same is not true of the nightly label printing and tracking number email process. This process can withstand a small number of errors and even be down for hours at a time without business impact. Though here, too, we can see the business stepping in, for example, only allowing the purchase of one label printer – or not keeping extra rolls of labels on hand.

Honestly though, one of the best business decisions I've seen a company make in recent years about reliability is a decision to not spend millions in building out a secondary system to swap over to when the primary is down. Why? Simple, the total downtime over the last few years was only a few hours every year – and the mechanism to switch over to the entire secondary system would take 30 minutes to complete. Given the response time and time it would take to determine you need to switch is at least another 30 minutes – you would be well into the outage a full hour before the secondary system would be active. So the business has to determine whether one or two more hours of uptime a year is worth the millions it would cost – they did not find value in that.

Mixing good and bad – tricks to wrapping bad code and making it resilient

As a rockstar SRE, I'll be the first to admit, I've put my fair share of duct tape in place to hold production systems up. You may think the job of SRE means we build out perfection, only do what is absolutely right or put together solutions that fix the root cause of issues – but you would be wrong. Our job, first and foremost, is to reduce revenue impact and protect the customer experience.

Alerting that fires actions

One of the simplest ways to provide corrective action is to build alerts that find the issue – then call scripts or actions to remediate it. The best example of this is actually built into most container base orchestrations, including Kubernetes. The infamous *liveliness* check Kubernetes makes to a container simply kills the container and spins up a new one when it fails. In short, if it doesn't respond when you poke it with a stick, it's dead – let's build a new one.

In today's world of logging and metric-based alerting systems, we can employ more advanced mechanisms. You can build alerts that tell you when one server in a cluster is experiencing errors in an application at a higher rate than the others and use a script to restart its service or application pool. Or flush a cache when we see new versions of applications rolling out. Or even tear down and rebuild new, bigger servers in production when we see the load increasing.

The sky is the limit, and while some may classify restarting an application pool or service as duct tape, we must consider the level of effort and lifespan of an application. Is the application due to be replaced in a few months – in this case, does it make sense to try to fix an application due for retirement? Is the problem impactful – like our infamous nightly label process at the custom hat company – does using duct tape matter if we simply restart a server at 3 a.m. to overcome a memory leak in an application?

Finally, sometimes, tracking down the root cause of an issue can take time, or the subject matter expert of a system or process is unavailable to remediate a production outage. Building a quick piece of duct tape to temporarily keep the application alive may be the best choice in these cases and, in the end, be the best way to bring a system back up or limit it until a formal fix can be put in place.

Just be mindful that some temporary fixes become formal fixes for years. This can also be okay if it's discussed as part of the SLA of a solution with business and technology partners. The important part is to understand how duct tape fixes can impact customers and revenue – the important part is always the discussion.

When we use alerting to remediate issues, it should be noted that the impact is already occurring to revenue and customer. It's a reaction and not a true fix.

Adding additional logging to monitor potential issues

One of the most effective tools an SRE has is visibility into how applications are running. And while metrics can provide great details into the quantities, it's the logs that provide the highest amount of detail. Adding logging lines into an application can be as simple as building a feature branch and asking developers to include it in their next release. When we identify an application with high failure rates or we just simply don't trust it, adding logging statements allows us to keep an eye on production.

The interesting thing about logging is that it is black and white; no longer do you have to go to the product team and say I have a feeling or I don't trust the development team's capability, or this third-party written application looks shaky – you can bring with you, undeniable, black and white proof that there are issues in the application. Most importantly, you can quantify these issues and bring together both revenue impact and customer experience stories. When you can say, for every one of these errors generating the pricing for a custom hat order, we lose a potential customer – and that these errors are happening 30 times an hour – you bring real visibility into the issue in a way the business can understand.

Using try catch to encapsulate exceptions

Few modern languages do not support the encapsulation of code in some type of a *try...catch*-type statement to capture exceptions and provide either alternative paths or even just skip some types of functionality. *Try...catch* is designed to wrap around lines of code, and when an exception occurs, your application doesn't crash; it simply captures the exception, then you can choose what happens next.

I always like to add logging to my *try...catch* statements to let me know when exceptions are happening. It should also be noted that certain **application performance monitoring** (**APM**) systems, will not show exceptions captured in *try...catch* and require an additional line of code to allow the APM to capture the exception when it happens.

And sometimes, we have code that isn't mission-critical to an application. A great example of this, oddly, is logging. Logging is rarely considered essential to an application – and I agree logging is highly valuable. Though, if I have a choice between servicing a customer and making revenue or getting my precious logging statements – it's no choice at all. I would caution that there are times when logging is essential, even required. There are many different processes where logging who has accessed information or that specific transactions are happening have a legal requirement to be tracked, and logging is one way of ensuring this happens.

Retries to the rescue…or not

There is much debate in the industry on whether or not to retry calls to systems and databases outside of a local program. Some feel if a system is failing, adding more calls to that system could overwhelm it even more, making it harder to recover. And for those of us who have seen systems overwhelmed and having an issue we are trying to recover from, that recovery issue is often considered a separate problem to be addressed. On the other side of the coin, when we talk about multiple containers or servers working through load balancers to respond to requests – if a container or server is down, a second or third call could make it to a different container or server and have a successful transaction.

So, where do rockstar SREs fall? In the middle.

For database calls, when a database is struggling, adding load may be the last thing you want to do – though, when databases are not well monitored, looking at retry calls and adding more load to bring the database down faster can be an effective last-ditch strategy used for calling attention to these types of issues and building logging trails to support additional need. However, this tactic isn't ideal and should be used with an abundance of caution.

For server and container-based applications behind a load balancer, there is an amount of time that these items must be unhealthy before they are taken offline – it is not uncommon for an unhealthy container to be in production for 1–5 minutes before it's terminated. Servers can be even worse. And when applications are in the flux of kind-of-working, if they fail only every other liveliness health check call, they can stay in production for days. To me, using retries on these types of architecture just makes good common sense.

When we look at applications that operate using newer function-based serverless orchestrations, such as AWS Lambda or Azure Functions, retries become even more of a gray area. One thing to look at is does the application use a third-party application that is server or container-based, and, if they do, does the application contain retries? If not, you may need to implement a retry strategy just to enable retries on the dependent third-party application.

And of course, I want logging when I have to retry issues – and this is where being a rockstar SRE comes to play. I don't want errors that are resolved with a retry to be logged like other errors. While understanding the number of retries happening in an application is important, when we look at raw error counts, errors resolved with retries should stand alone. Why? The biggest reason is that if errors resolved with retries are counted as normal errors, you get a false sense of the application's ability to fulfill its primary purpose, generate revenue, and service customers. Remember, these are still happening if a retry resolves the initial error. And most importantly, if this happens at 3 a.m. I do not want to wake my staff because retries are working as they should to resolve errors. Though, I'd want to know when I got in the office the next day if it was excessive.

Summary

I often say that revenue is made somewhere between a perfect system and a complete ball of duct tape. Any book on entrepreneurship and business growth will tell you there is risk in everything we do. Understanding how much you're willing to risk and how to mitigate the risk are the backbone of growth. As SREs we can leverage observability to track the possibility of risk and know when a system or application may be slipping to provide revenue. We can further use tools and even duct tape to mitigate that risk in code and architecture. It's not about perfection; it's about the best decision to keep revenue flowing and keep enough of your customer base happy to maintain a growing business.

There are multiple ways business needs impact the quality of code and architecture – understanding these can help you perhaps change some of these events, and when you can't, you can often combat these issues with higher levels of observability and more deep dives into the code and development process. Of course, you'll never mitigate it all – but understanding the cause helps us better prepare for possible disasters.

We've talked much about customer and business impact in these two chapters – just keep in mind, when we talk about business – those, along with costs, are the key indicators that should be driving decisions.

In the next chapter, we will discuss the essentials of **metrics, events, logs, and traces (MELT)**. Or as I like to call it, the basics of observability. Most importantly, we'll explore how these different views of the system are best leveraged by the rockstar SRE.

Part 2 - Implementing Observability for Site Reliability Engineering

The second part focuses on the skills and knowledge any **site reliability engineer** (SRE) must have to succeed in this profession. It explains what a SRE should learn about observability, systems administration, and how data science is part of their day. Through the chapters in this part, the reader will find comparison tables that will elucidate how the site reliability engineering profession differs from other IT professions and how to find more information to hone those skills.

The following chapters will be covered in this section:

- *Chapter 4, Essential Observability – Metrics, Events, Logs, and Traces (MELT)*
- *Chapter 5, Resolution Path – Master Troubleshooting*
- *Chapter 6, Operational Framework – Managing Infrastructure and Systems*
- *Chapter 7, Data Consumed – Observability Data Science*

4

Essential Observability – Metrics, Events, Logs, and Traces (MELT)

If you cannot find any problems, then none exist! Are you sure about that? You probably wish this was true. However, in the systems visibility context, we can assume if we don't find any problems, most likely, we have a blind spot. To illustrate that, how many times have an operations team heard about service degradation from its user first?

Observability, in making systems *observable*, is a notable feature of site reliability engineering but also one of its guiding principles. We can define it as the means of understanding the inner states of a system (or solution) by inspecting its outputs or signals. Although observability is an evolved telemetry model, where we want to collect measurements from a distance to minimally affect the measured system, it may require logging and tracing components to sharpen the system visibility. Observability has many characteristics, techniques, and considerations that we will explore in this chapter, along with a simulation lab to gain practical knowledge.

> **Important note**
> Recently, the market adopted the term "o11y" to refer to observability. You may find observability documentation, tools, and courses by searching for o11y, which we pronounce "ollie." From the site reliability engineering point of view, observability is necessary but not enough. Giving a memorable name to one of the reliability pillars is cute, but we should not make it more significant than the others.

In this chapter, we will explore the basic aspects of systems monitoring and telemetry by explaining what they are and why they are key for observability. We will provide a good understanding of how SREs monitor applications nowadays. After the basics have been covered, we will link monitoring back to observability and clarify why they are not the same thing. Also, we will step into alerting and notification by explaining why they are important and how to do them in the site reliability engineering style.

In this chapter, we're going to cover the following main topics:

- Accomplishing systems monitoring and telemetry

- Understanding **application performance monitoring** (**APM**)

- Getting to know topology self-discovery, the blast radius, predictability, and correlation

- Alerting – the art of doing it quietly

- Mixing everything into observability

- In practice – applying what you have learned

Technical requirements

Deploying a monitoring platform that conforms to the observability tenet is not difficult. Since we wanted accessibility over wider features, we selected open source software in most cases for this monitoring system. When open source software was not available, we have chosen a software that has a free offer so that anyone can gain practical knowledge from this chapter.

You will need the following for the lab:

- A laptop with access to the internet

- An account on a cloud service provider (we recommend **Google Cloud Platform** (**GCP**))

- The kubectl CLI tool installed on your laptop

All manifest files, scripts, and application code can be found on the public GitHub at https://github.com/PacktPublishing/Becoming-a-Rockstar-SRE/blob/main/Chapter04.

Accomplishing systems monitoring and telemetry

Monitoring IT systems is undoubtedly a must-have requirement for any solution or service. It started a long time ago with the Unix operating system. If you are old enough, you may have used tools such as **top**, **vmstat**, or **syslog** to monitor critical processes and their performance and usage of resources. We *monitor* to watch the utilization and responsiveness of IT resources (or services) to determine whether the system running on top of them is available and performing under nominal parameters. A lot has changed since then as systems have become more complex and distributed. Monitoring expanded from locally executed tooling to a remote collection of measurements, also called **telemetry**, to minimize its own footprint (or consumption of resources). We will divide this topic into four sections:

- Monitoring targets for infrastructure

- Monitoring types and tools

- Monitoring golden signals

- Monitoring data

Let's begin with understanding what is monitored in a system.

Monitoring targets for infrastructure

In a typical system that is composed of infrastructure and applications, indispensable IT resources are required by the application tier to run according to its design. SREs are responsible for managing these resources, so they need to ensure the monitoring platform is surveilling them. In other words, SREs are directly responsible for the effectiveness of **observability**. We have, at the bare minimum, the following critical resources in any solution: the compute, network, storage, database, and middleware. We are going to go through each of them now.

The compute

This is probably the IT asset that has transformed the most over time. We started with physical servers, for which we had to acquire the dedicated hardware, plug it into a data center rack, and turn the power on. Although this is rarer and rarer to see, we still have cases like this. In those scenarios, we need to monitor the hardware components and the operating system running on the server. Many hardware components have sensors and provide measurements to the hardware monitoring software, such as CPU temperature, fan speed, and CMOS battery voltage. At the operating system level, monitoring tools can provide a detailed view of the server resources, including CPU load, disk utilization, I/O transactions, vital processes, processes running, threads per process, and memory consumption.

As computing resources have evolved, we now have **virtual machines** (**VMs**) commonly used by many systems nowadays. VMs are nothing more than processes running inside a host machine that are controlled and managed by a hypervisor. Those processes simulate a virtual server that has everything a bare-metal server has, except for the hardware piece. In this case, we don't monitor the hardware layer as this is the responsibility of the cloud service provider. If we are running VMs in a local data center, then we need to monitor the hypervisor itself and the host operating system as well as the VMs themselves. The VMs are monitored like any operating system.

If we go further in the cloud adoption model, then we have containers as the compute resource. Containers are virtualized processes that have not just a virtual operating system and hardware but also additional system and application libraries. They are orchestrated by a tool such as **Kubernetes (K8s)**. The way we monitor containers is radically different from VMs and physical servers. Because they are made for distributed computing, where the load is chunked into small threads and each thread is processed by a different container, we usually monitor the total amount of CPU load, memory usage, or other resources, utilized by all containers together.

> **Important note**
>
> Serverless is the next step in the cloud adoption ladder. It's important to note that we don't monitor a serverless deployment compute resource. This is entirely the responsibility of the cloud service provider (also known as the **hyperscaler**), yet we do need to monitor the application running on it.

The network

This is the group of IT resources that is the hardest to make visible due to its complexity. First, network devices have various distinct network functions and can be based either on white- or black-box hardware. Because networking always been distinct from the rest of IT technologies, it's not a surprise when we hear of something like **network reliability engineering (NRE)**. Although we agree it is an intriguing idea to have a distinct SRE specialty for networks, with the advent of **software-defined networking (SDN)**, it seems it may not stick at all. Thus, Jeremy and I consider NRE already a part of SRE in this book.

We can start by talking about the different types of network devices that may exist. Among them, we have routers, switches, hubs, bridges, gateways, and repeaters. They can be either physical or emulated devices. Usually, they have **Simple Network Management Protocol (SNMP)** enabled so device monitoring can be accomplished by using SNMP traps to receive important events from these devices.

Networking uses a multitude of network protocols to interconnect devices on multiple layers. Looking at the **Open Systems Interconnection (OSI)** model, also known as ISO/IEC 7498, you can see seven layers including the physical, data link, network, transport, session, presentation, and application layer (see *Figure 4.1*). For each layer, there are distinct protocols depending on the manufacturer and the type of connection. That poses a large combinatory problem for the network monitoring systems, which need to handle a great number of possibilities:

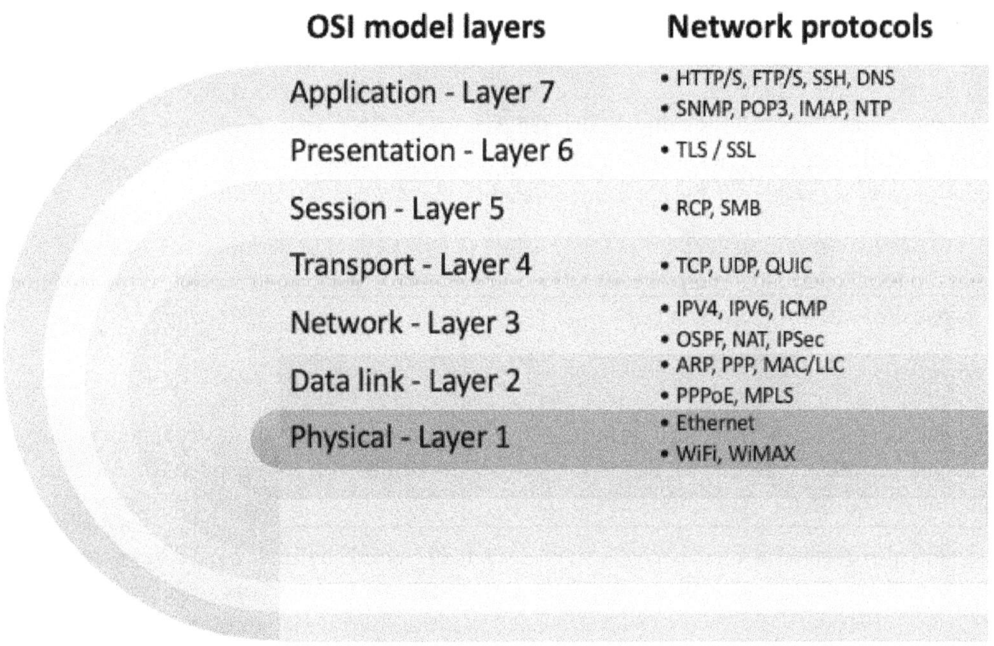

OSI model layers / Network protocols

OSI model layers	Network protocols
Application - Layer 7	• HTTP/S, FTP/S, SSH, DNS • SNMP, POP3, IMAP, NTP
Presentation - Layer 6	• TLS / SSL
Session - Layer 5	• RCP, SMB
Transport - Layer 4	• TCP, UDP, QUIC
Network - Layer 3	• IPV4, IPV6, ICMP • OSPF, NAT, IPSec
Data link - Layer 2	• ARP, PPP, MAC/LLC • PPPoE, MPLS
Physical - Layer 1	• Ethernet • WiFi, WiMAX

Figure 4.1 – Example of existing network protocols per OSI layer

For that reason, most network equipment already has monitoring capabilities embedded within it. SNMP is a *de facto* industry standard, and it can also be used to capture network telemetry data.

Over time, network monitoring has moved from simply checking network devices and the protocols that make networking possible for them to a more scalable, robust, multi-vendor, and elegant capacity called **network performance monitoring** (**NPM**). Under this concept, these tools provide very interesting features, to name a few:

- **Wi-Fi or 4G or 5G heat mapping**: This analyzes the Wi-Fi or 4G or 5G signal intensity and power for an area.

- **Network device scanning**: This discovers the network devices in a determined subnet using SNMP (or through other means) and makes sure they are monitored.

- **Network performance testing**: This involves continuously testing the network performance, availability, and latency to identify degradation or faults. It allows a more proactive stance for large networks.

- **Network diagnostic tooling**: This provides visual tools with analytics to help network engineers or SREs troubleshoot network performance issues.

As with compute resources, virtualization has reached the network. Many network hardware appliances such as routers or load balancers are available as virtualized devices or processes running inside a VM. They can even run as containers. This is called **network functions virtualization (NFV)**. A parallel virtualization technology effort is **software-defined networks (SDN)**, where a software-based network controller directs traffic on a network. For instance, on an SDN, routing and firewalling are done by the control plane software over the application data flowing through this network.

NFV and SDN are commonplace nowadays, mainly in hybrid cloud environments. SREs need to understand those technologies and work with a proper network performance monitoring tool that also targets these network technologies.

Storage

Unless you are currently working in a data center, you won't see actual storage devices often. Probably one of the first resource groups that was virtualized and revamped, it is mostly offered as a service to applications by on-premises or cloud infrastructure.

Storage has also evolved at a fast pace, like networks. It all started as simple hardware components (called **hard drive disks**) inside microcomputers where persistent memory was a requirement. Later, they became appliances of their own with dedicated hardware and embedded software (**firmware**). In the beginning, we had physical disks. Later, those disks turned into virtual disks based on array formations of physical disks (**Redundant Array of Independent Disks (RAID)** technology), and now the storage capability is a software object inside a cloud platform at an application's disposal due to extreme virtualization.

Storage monitoring has evolved from simply asking how much space is left for applications to storage performance monitoring, where it assesses the following items:

- Current capacity, present consumption, and future projections
- Device ports, channels, controllers, physical disks, and disk group health and performance
- Physical disk, disk array, and disk group **input/output (I/O)** throughput
- Compute hosts' demand and request

> **Important note**
> Compute, network, and storage are certainly the most virtualized resources, with multiple native management functions. However, that doesn't necessarily mean good performance is guaranteed. Systems fail and have slowdowns or downtime; monitoring is a must even for virtualized environments.

Databases

Databases hold the data of the world! Either a **relational database management system (RDBMS)** or a non-relational database management system, they are the resources responsible for handling and managing data structures and models. Most applications use one or more databases to store and retrieve data through queries. It's common to adopt a NoSQL database, which is another term for a non-relational database.

We deploy an RDBMS or NoSQL DB to serve an application in various ways. They can be installed on a VM, deployed as containers, or consumed as a service in a cloud environment (also known as **database as a service (DBaaS)**). As another pivotal IT resource, we need to make sure all databases are in the monitoring targets' roster.

Database monitoring tools should evaluate the following:

- **SQL statements**: An RDBMS uses **Structured Query Language (SQL)** to record and retrieve data from tables. These SQL statements are expanded to plans, and there is a computing cost to calculating the dataset to retrieve and reading it from storage. Slow statements can lead to unresponsive systems.

- **Database locks**: An RDBMS may receive multiple SQL statements to change the same dataset. Every time a dataset change is requested, a DB lock happens to ensure no other process can change it. Long locks are very expensive and may make a system unreliable.

- **Database performance**: This is gauged through the continuous evaluation of the health and traffic of database queries, as well as how the underlying storage affects DB operations.

Middleware

Middleware software provides a neutral application runtime platform that enables application portability while abstracting the underlying platform and network technologies from application developers. Middleware resources are known for their diversity. There are dozens of vendors and hundreds of distinct software products and services. One way to categorize them is as follows:

- **Application platforms**: This segment contains application runtimes and the platforms on which the business logic and presentation tiers run, including technologies such as application and web servers.

- **Hybrid integration**: This segment contains **enterprise application integration** (EAI) technologies such as message-oriented middleware, business process management, and **business-to-business (B2B)** middleware products.

- **Application security**: This segment contains application identity management tools such as directory servers and access managers.

- **Big data analytics**: This segment refers to middleware products responsible for big data analytics engines and infrastructure. It includes large data distributed processing and unified analytics tools.

- **Engineered systems**: This segment contains middleware solutions based on a combination of specialized hardware with middleware software embedded into it. It spans from simple single-function appliances (usually 1U-7U rackmount in size) to complex multi-function integrated systems (usually taking up a whole rack).

- **DevOps tooling**: This segment contains middleware products used as part of the DevOps delivery pipeline infrastructure.

It's not a surprise that middleware software products have been ported to cloud services providers using a **software-as-a-service (SaaS)** model. In this scenario, the **hyperscaler** is responsible for managing and monitoring the middleware.

Each middleware category has specific monitoring requirements. We advise checking the technical specifications for each product to see the monitoring recommendations. Moreover, there are a few monitoring tools capable of watching middleware functions and services. In those cases, you probably need to use a custom watchdog script.

We have covered all the monitoring targets for the infrastructure layer – we will look at the application tier in detail in the *APM* section.

Monitoring types and tools

There is a multitude of components that we should monitor as SREs. To accommodate this diversity of targets, we have many monitoring types and tools. There is no consensus around the monitoring taxonomy among vendors, as each of them may have distinct terms. We provide our own classification for the sake of learning here.

Monitoring types

We consider five IT monitoring types: availability monitoring, **APM**, **application programming interface (API)** monitoring, synthetic user monitoring, real user monitoring, and **business activity monitoring (BAM)**. Let's check each of them.

Availability monitoring

This is the most common monitoring type, also called system or **IT infrastructure monitoring (ITIM)**. It continually checks whether a system, including its infrastructure components and services, is available to its users. Availability is not the same as uptime. A system may be available but not operating within a minimum level of performance. In such cases, where there is service degradation or slowness, the system is considered to be in downtime.

This type is responsible for watching the IT resources that we discussed previously such as the compute, storage, networks, databases, and middleware.

APM

This type of monitoring evaluates an application from the user's perspective. It supplies a good detailed level of information for spotting issues related to the application performance. We are going to dig into APM in the next section.

API monitoring

APIs are very common in applications using the microservices architecture pattern. They provide services to other applications through a web interface. This monitoring type ensures those APIs are working as expected in terms of performance and data accuracy.

User monitoring

Synthetic user monitoring relies on **bot** technology or other end user mimicking techniques to simulate user interaction with the application or web service. This type of monitoring provides more accurate data about the performance of the application functionality, as it tries the complete interaction flow.

Real user monitoring (RUM), or **digital experience monitoring (DEM)**, tracks the user experience inside the application and through its interactions with the application **user interface (UI)**.

BAM

This is the most interesting monitoring type for businesses. It's responsible for assessing business **key performance indicators (KPIs)** based on the IT monitoring systems. BAM cannot work without the previous monitoring types, as it's built on top of them. Also, the business activity must be mapped to the underlying IT processes, systems, and transactions – for instance, how many sales happen on an e-commerce portal over time.

> **Important note**
> Although we don't detail BAM in this book, SREs should have a basic idea of how business activities are related to business metrics. Some of the most common business metrics are average revenue per user, the average cost per user, total addressable market percentage, user churn rate, net promoter score, the average cost of acquisition per user, and average user lifetime value.

Monitoring tools

There are two deployment architectures for monitoring tools: **agent-based** and **agentless**. Agent-based ITIM tools require that a specific agent is installed and configured in the monitored entity. On the other hand, an agentless monitoring tool makes use of existing venues to collect data from targets.

Agentless monitoring

The name **agentless** is debatable, as it doesn't signify that monitoring tools in this category can read telemetry data out of thin air. They use existing management protocols to gather monitoring data from targets. Examples of these protocols are SNMP and SSH. The inherent limitation of agentless monitoring is the fact that all targets need to have these protocols enabled.

Agent-based monitoring

In this category, we need to install an agent (a tiny footprint software) on the target. The monitoring tool communicates with each agent to read monitoring data and inject commands into remote systems. This gives the monitoring tool much greater capabilities; however, it brings additional costs in terms of licensing and the maintenance of the monitoring platform. As a limitation, the agent is available to a specific set of systems.

Besides those two deployment architectures, we can categorize monitoring tools depending on their purpose. We count five categories here: observational, analytical, event and log management, distributed tracing, and visualization tools.

Observational tools

As the name indicates, these are the most basic monitoring tools, which observe IT systems, infrastructure, services, and resources. Almost all monitoring types and categories fall under this category. Usually, APM, synthetic user, and real user monitoring types have built-in analysis capabilities, so they go beyond just observation. A few examples of observational tools on the market are Grafana Mimir, Nagios, ScienceLogic SL1, IBM Tivoli Monitoring, and Prometheus.

Analytical tools

Those tools further analyze observational data and combine it with other data sources to detect anomalies and their possible causes in real time. It may also apply **artificial intelligence** (**AI**) models to uncover patterns and predict problems. A few examples of analytical tools on the market are Dynatrace, AppDynamics, Datadog, Splunk, and BigPanda.

Event management tools

Event management is part of the **IT Infrastructure Library** (**ITIL**) standard. It demands that all IT monitoring data should be treated as events. By collecting, standardizing, and consolidating monitoring data from all sources into events, further data analytics is possible, including the application of machine learning models and AI features. Most analytical tools transform ingested data into events. A few examples of event management tools on the market are IBM Tivoli Netcool, Splunk IT Service Intelligence, SolarWinds SEM, and ServiceNow ITOM.

Log management tools

IT infrastructure, services, and applications produce logs through their logging process. These logs contain important information about a segment of the performance, state change, user activities, usage, billing, and security. A good monitoring system must ingest and aggregate these logs to analyze the combined data. A few examples of log management tools on the market are Dynatrace, Splunk, Fluentd, Logstash, and Grafana Loki.

Visualization tools

There's no best way to make sense of monitoring data, which can be considered big data, and then visualize it as graphs, gauges, and counters with innate statistical models all together in dashboards. These tools help SREs learn system patterns and anomalies to improve monitoring (and consequently observability). A couple of examples of the visualization tools on the market are Grafana and Kibana.

Distributed tracing tools

Maybe the latest category to emerge in the monitoring domain, distributed tracing tools have been created for microservices-based applications. These applications have numerous moving parts called microservices, which can be used in various ways. Determining which microservices and components have been used for a certain user request may be unmanageable if there are dozens of possibilities. To tackle this problem, applications enable tracing a user's request through the utilization of special headers. Based on these headers, it's possible to track which parts of a system served a specific user operation. A few examples of distributed tracing tools on the market are Grafana Tempo, OpenTelemetry, Jaeger, Dynatrace, New Relic, Instana, Datadog, and Splunk.

We see an increasing difficulty in categorizing modern monitoring tools as more and more, they incorporate new features and release sub-products. There is a clear trend of a single software vendor having all the monitoring capabilities that were covered by multiple vendors in the past. Anyway, the intention of this taxonomy is merely to explain the main monitoring functionalities to the reader.

We find this debate about which monitoring suite is better than others endless. As SREs, we know better than anyone else that there's no one-model-fits-all approach. A structured decision-making process must be used to determine which monitoring types and categories should be adopted and why.

Now that you understand the monitoring taxonomy, let's explore the famous golden signals.

Monitoring golden signals

Signals in our context are pieces of information measured by a monitoring system at the monitored target to convey vital characteristics about it. Google defined *The Four Golden Signals* in their book. They coined the term *golden* to implicitly say these four signals are the most meaningful monitoring aspects for any user-facing system or sub-system. According to Google's book, the four golden signals are **latency, errors, traffic, and saturation (LETS)**. We use an acronym to make them memorable.

Latency, Errors, Traffic, and Saturation (LETS)

This is the original Google version, with one tiny modification in terms of the order to create this acronym:

- **Latency**: Latency is the time it takes to respond to a user request. Imagine that you access an internet banking portal. The first thing you're going to realize is how many seconds it takes to show you the portal after you click on the button. This is a KPI for systems that serve users. Any degradation or slowness will immediately affect this indicator. Also, errors such as HTTP 404 (page not found) have much less latency than a valid page. Differentiating latency for good and bad responses is required.

- **Errors**: We don't want to know the number of errors the monitored system users receive, but rather the ratio of errors (HTTP 50x codes) and success (HTTP 20x) responses. That means this signal presents a percentage of successful responses against the total requests and, for that reason, it's called an **error rate** signal sometimes. Also, failure is not limited to undesirable HTTP codes. You can define conditions under which a response is considered a failure. For instance, if a successful response is sent to the user after 1 minute, it's treated like an error.

- **Traffic**: This signal has a heavy dependency on the target type. Traffic has distinct connotations depending on the system or sub-system. For web-based user interfaces, the number of HTTP requests per second is a good example of this golden signal. It translates directly to how busy your system is and how many activities are happening per unit of time. Sometimes, this is called **utilization** as well.

- **Saturation**: This is a measurement of the current load for a system. The trick here is to know your breakpoint based on the IT resources available. If your system is constrained by the CPU, then you don't want to reach 100% of its utilization, as this would cause service degradation for sure. Therefore, the **saturation** signal is based on the CPU load. However, modern applications live on *hyperscalers* where auto-scaling is possible. In those cases, network bandwidth is a constraint, so the saturation indicator is based on how much of the available bandwidth is consumed.

Saturation, traffic, error rate, latency, and availability (STELA)

STELA adds one more golden signal – **availability**. Although the availability of a system can be determined by any of the four golden signals, many SREs think this is a good practice. You start by testing the availability signal and if that fails, then you don't measure the other golden signals, as that would be a waste of resources.

Utilization, saturation, and errors (USE)

The USE method is a methodology created by Brendan Gregg that uses checklists to assess system performance. It's very interesting for monitoring IT resources, especially compute nodes.

Rate, errors, and duration (RED)

The RED method is a methodology created by Tom Wilkie that defines three metrics that should be measured on all microservices in the environment. The monitoring metrics defined by this method are as follows:

- **Request rate**: The number of requests per second
- **Request errors**: The number of unsuccessful requests per second
- **Request duration**: The time each request takes per phase

Although these methods are not based on the golden signals, they are correlated. The principle of observability starts with defining the right spot to place monitoring measurements. SREs rely on LETS, STELA, RED, and USE for that. Next, we will examine what kind of data monitoring systems create, store, manipulate, and manage.

Monitoring data

Another way to view monitoring is through the data it handles. **Metrics, events, logs, and traces** (also known as **MELT**) are the heart of the observability tenet. The quality of this dataset decides how effective the monitoring platform is at making a system observable.

Metrics

Metrics were the first data type created by monitoring tools and they are the basis for accomplishing observability. Metrics are aggregations of the monitoring measurements made at regular intervals. They hold a value for a specific aspect of the target system or component – for instance, the number of requests served by a web application to its users. Another example is the rate of successful operations in the last hour.

Most monitoring tools save metrics in a typical and optimized format, which includes a timestamp, a name, a few labels (dimensions), and a value. **Prometheus** (an open source monitoring tool) stores each metric as distinct time-series streams and when retrieved via its API, the metric data appears like this:

```
{
    "status": "success",
    "data": {
        "resultType": "matrix",
        "result": [
            {
                "metric": {
                    "__name__": "http_request_duration_seconds_
sum",
```

```
              "code": "200",
              "method": "GET"
       },
       "values": [
           [
                   1662073553.318,
                   "0.446390"
           ],
           [
                   1662073553.282,
                   "0.4301564"
           ], ...
       ]
   }
  ]
 }
}
```

The first highlighted field is the metric name. The next two fields are labels, also dimensions of this matrix, as you consult columns individually. Then, you see pairs of the timestamp and value measured at that point in time. By keeping historical data for each metric, it's possible to apply statistical models to obtain a deeper analysis.

The golden signals cited in the previous section are monitoring metrics. More examples of metrics using the golden signals are as follows:

- **Latency**: The time in seconds that 99% of the requests are served

- **Error**: The percentage of responses with HTTP code >= 500 in the last 10 min and the percentage of responses with HTTP code 2xx in the last 10 min

- **Utilization (traffic)**: The number of HTTP GET requests per second

- **Saturation**: The number of HTTP requests with a waiting status

Metric **thresholds** are conditions that when satisfied, trigger an event. For instance, if the proportion of average error codes is greater than 10%, an event is distributed to notify you about this happening.

Observability cannot be reached if metrics are not well defined, collected, and analyzed. Metrics should be gathered continuously and without interruption, meaning the monitoring configuration must be reviewed often.

Events

IT systems, applications, services, and infrastructure are subject to constant changes. These changes are discrete in nature and don't require specific monitoring. Instead, events are created to communicate what has happened at a high level and when exactly it has happened. For instance, a new application release has been deployed to the production environment. This occurrence triggers an event that is distributed throughout the monitoring system. Now, it's possible to contextualize all other monitoring data considering this event.

Like metrics, we can add metadata such as labels to enrich event information. A common practice is to include the configuration item or components in this information. That way, it's possible to determine the sub-systems involved in that event.

Even monitoring metrics can trigger events based on **service-level indicators** (SLIs). It's not rare to see a monitoring system where MELT data is reduced to events and a single structured data source is provided to the analytical tools.

It's common to see events in cloud environments to audit service object creation, deletion, and updates. This is another use case for events, other than the ones already discussed.

Logs

Applications and other IT resources spawn an immense quantity of logs. Logs are low-level detailed information on events that occur inside systems. They carry several contextual fields to help SREs troubleshoot an issue for a specific sub-system. With this level of detail, it's possible to play back what happened to an affected component. For instance, this is a K8s informational log:

```
I1025 00:15:15.525108          1 httplog.go:79] GET /api/v1/
namespaces/kube-system/pods/metrics-server-v0.3.1-57c75779f-
9p8wg: (1.512ms) 200 [pod_nanny/v0.0.0 (linux/amd64)
kubernetes/$Format 10.56.1.19:51756]
```

Notice the quantity of information you can get in a single log entry. The first field is the log level and code. Then, there's a **timestamp** and the **pod** name.

Each product may have a particular log format, but sometimes those log entries are not normalized, which leads to unstructured data that is hard to analyze. For that reason, log management tools can be used to normalize, aggregate, and analyze log data.

Due to the possible quantity of log entries, log aggregation becomes a data lake problem that must be resolved with the utilization of big data analytics technology.

Logging is the process of creating log entries. Each log entry is assigned a level depending on its criticality. Each software vendor may define its own scale. We have opted to refer to **Log4j**, which is the most widely adopted logging library. Here, we list the levels of criticality for logs according to the Apache standard:

- **Fatal**: A fatal event that provokes an outage for an application.
- **Error**: A error in an application – it's possible to recover from it.
- **Warn**: An event that may turn into an error.
- **Info**: An informational event.
- **Debug**: A general diagnostic information event.
- **Trace**: Fine-grained diagnostic information that captures where in the code the event happened. Do not confuse this one with traces as a data type from MELT.

You can find more information at this link: `https://logging.apache.org/log4j/2.0/log4j-api/apidocs/org/apache/logging/log4j/Level.html`.

Traces

Distributed tracing tools are the latest ones to surge in the monitoring domain. A trace is a chain of correlated events belonging to the same application. Distributed applications that are built using a microservices pattern work as a service mesh with multiple small components called microservices. Each application transaction can use a unique combination of available microservices. Traces identify the set of microservices used for a certain transaction through header propagation using a transaction ID.

The following is an example of an HTTP header inserted by an application to indicate a trace:

```
X-Trace-Id: "4c048ed6-3c6f-4f7f-881f-39219bbc2e75"
```

You can find more information at this link: `https://www.w3.org/TR/trace-context/#traceparent-header`.

A good monitoring system has all the MELT data described here. That makes a system observable from all angles. Also, the consolidation of this data into a single data lake supports the application of more advanced analytics and AI models. However, the data quality should also be considered. We like to say that if your data items are easy to understand, detailed enough, labeled accordingly, and live for long enough, then they form a good dataset.

We have summarized systems monitoring, so now we will jump into the APM specifics.

Understanding APM

We want to expand on APM a little further, as applications (apps) are the gravity center of any system. Traditionally, monitoring always started at the infrastructure stack, while the application was a lesser concern for checking metrics. That changed radically in the last two decades. After the pandemic of 2020, APM is now at the core of the digital transformation that the world is experiencing.

APM is now a synonym for unified or all-in-one monitoring. It's the culmination of all monitoring types and tools. It offers an end-to-end monitoring capacity with additional automation, alerting, and analytical capabilities.

According to Gartner in their Magic Quadrant for Application Performance Monitoring and Observability research, they defined APM "*as software that enables the observation and analysis of application health, performance and user experience. The targeted roles are IT operations, site reliability engineers, cloud and platform ops, application developers, and product owners. These solutions may be offered for self-hosted deployments; as vendor-managed, hosted environments; or via software as a service (SaaS).*"

The features of APM may differ from vendor to vendor, but some are very common to most forms. Let's check them out:

- APM
- Microservices/API monitoring
- RUM
- Server/compute/serverless monitoring
- Database/data service monitoring
- Network monitoring
- Cloud monitoring
- Security monitoring
- Application instrumentation libraries
- Data visualization and analysis

Another way to describe APM suites is that they are the monitoring technology that makes MELT data collection possible and useful for SREs.

Monitoring should always start at the application functionality tier and move toward to infrastructure layer. For that reason, APM is key to keeping an eye on this functionality, whether the user experience, application performance, or infrastructure metrics. We are going to discuss the latest monitoring technology trends in the market next.

Getting to know topology self-discovery, the blast radius, predictability, and correlation

This is where monitoring gets interesting. With normalized and combined MELT data, we can use more advanced data science techniques to uncover what's happening inside a system based on certain patterns in a dataset.

AI models are becoming a standard feature offered in most APM suites. An AI model is nothing more than specialized software built with artificial neural networks. Those algorithms can be trained to recognize patterns, including anomalies. If what characterizes normal and abnormal behavior is understood by the AI model, it can predict failures before they impact the user by checking whether the behavior is trending in a certain direction.

AI models ingest MELT data to do machine learning. This first set of data is called training data, which tells the AI model what's good. Then, it creates a hypothetical model of the data to find patterns inside it. The next step is to test the AI model against real data to determine whether it detects abnormal conditions.

This kind of approach to monitoring data enables something called AIOps, meaning SREs can operate a system using AI assistance. The most meaningful AI-enabled features are as follows:

- **Events correlation**: Events can be correlated by grouping them using certain patterns and rules. The simplest model is to group duplicated events such as a metric threshold that has been reached more than one time for the same configuration item. Grouping repeated events is also known as deduplicating events and it assists with decreasing event noise. A more complex scenario is grouping events for the same application transaction passing through multiple infrastructure elements and cloud providers.

- **Anomaly detection (predictability)**: Once the correlation of MELT data types is in place, the AI model can extract patterns from a dataset. It can decide whether certain patterns should be handled as anomalous candidates and engage SREs for further analysis. Since those anomalies can be detected earlier in the process, many times an automaton or human actor can intervene to alleviate the root cause and avoid an incident. The ability to detect problems before they start to cause heavy symptoms is called predictability. Many companies brag that they provide this feature in their suites. However, without full-fledged SREs selecting which SLIs and MELT should be prioritized, the monitoring data may not be good enough for predictability.

- **Topology self-discovery**: By having metadata such as labels and tags added to MELT data, it's possible to reverse-engineer and ascertain the system topology by learning about the relationships between components. Topologies are extremely useful for troubleshooting problems that affect more than one sub-system.

- **Blast radius**: With the solution topology at hand, the AI model can determine which components have been affected or are affected by an anomaly. This information helps SREs to better isolate the causes of service outages or slowdowns.

A monitoring effectiveness indicator is how fast an anomaly is detected. If you can identify a still-building-up problem before it impacts the system user, you're on the right path. Next up, let's examine what happens after monitoring captures an issue – it sends out an alert to the SREs.

Alerting – the art of doing it quietly

Since many of monitoring tools have alerting functionalities, we want to talk about how SREs define and handle alerts.

First, let's understand what an alert is. As we have learned, modern monitoring systems work with events – normalized and structured monitoring data types. Some of those events are considered critical and urgent. When that happens, the monitoring system needs to raise an alert that goes to a notification system. The notification system lets the first responders know about the alert.

Alerting and notification are straightforward processes but moreover, they need to be cost-effective. As SREs, we should pay attention to a few guidelines to ensure that the outcomes of these processes add value to the end user rather than giving us more operational work. For that purpose, we will divide this topic into two sections.

The user perspective notification trigger principle

This SRE principle advises us on how we answer the question, what is critical and urgent enough to turn an ordinary event into an alert?

The short response to this dilemma is simple – does this event directly affect the service performance *experienced* by its user? If the reply is *yes*, then most likely it's a case for an alert. If the answer is *no*, then more investigation is required.

The principle states that an alert should only be triggered if the end user experiences symptoms of the event. For instance, IT resources may be fully utilized, yet the system responsiveness is below the expected threshold. This is an event for sure, as we need to check why we lack additional resources. However, it's still not a problem for the users. They are happy.

On monitoring systems in which predictability is implemented, this principle needs a second condition. Trigger an alert if an event will soon lead to a system disruption or degradation from the user's perspective. There's no immediate impact on the end users, but the AI model is telling us there'll be one shortly with a given percentage of accuracy (usually over 80% for well-trained models).

Event-to-incident mapping principle

This is not an SRE principle, but an **Infrastructure Technology Service Management (ITSM)** principle. Events of any kind that may reduce the quality of service for a system should generate an incident. That means incidents are raised for events that are connected to any service disruption or performance issue. The reason for that is to track outages and downtimes for any component or configuration item. That track will feed the **mean time to detect (MTTD)**, **mean time to repair (MTTR)**, and **mean time between failure (MTBF)** metrics.

Incidents have levels of severity depending on the system impact they sustain. Like alerts, incidents can be defined based on the potential impact they may have on the system if ignored. The levels are **Severity-1 (S1)** (or **Priority-1 (P1)**) to S4 (or P4), with the most critical and severe level being S1 or P1.

To optimize the alerting process, alerts should come directly from the monitoring platform and not from the incident ticketing system. Nevertheless, alerts should always be documented as S1 or S2 incident tickets to keep track of incident management times and durations.

We just saw how alerting and notification are accomplished using SRE principles. Let's check how we bring everything together into observability.

Mixing everything into observability

One of the most interesting discussions on observability is how it differs from monitoring or even if it's different in any way. Monitoring does not depend on observability; it can exist and has existed without the objective of making a system observable. You can monitor anything just for the sake of watching it. Nevertheless, you cannot fulfill observability without monitoring.

Observability is much more of a principle than its main technology pillar: monitoring. Monitoring vendors, products, and technologies will change over time, while this tenet will not. We like the idea that observability is the right mix of monitoring types, tooling, and data to bring the inner states of a system to the surface to determine its behavior patterns and predict the next problem.

With this idea in mind, we want to discuss other aspects of this principle. We will split this topic into three sections.

Outages versus downtime

Let's clarify the misuse of some terms. It's essential to note that the meaning of **service availability** had changed. As monitoring precision and the **service-level objective (SLO)** targets have increased, service availability has moved from a binary (up or down) to a scaled (based on performance) status.

In the past, we had two possible service states, either available or unavailable. For that, our SLOs were based almost exclusively on availability metrics. Now, with other types of metrics, it's possible to measure service performance. How well the system is serving the users? If the performance is under the minimum value, the service should not be considered available.

The following terminology addresses these clarifications:

- **Available time**: This is the overall duration (sum of time) for which a service or system was running and available.

- **Outage time**: This is the overall duration (sum of time) for which a service or system was completely unavailable.

- **Downtime**: This is the overall duration (sum of time) for which a service or system was unavailable or performing below the minimum rate.

- **Uptime**: This is the total time minus the downtime.

- **Reliable time**: This is the difference between uptime and downtime.

Now that we have clarified those concepts, let's look at the observability architecture.

Observability architecture

We are not in favor of using the terms APM and observability interchangeably. We think that observability, as an intention, should be technology-agnostic as much as possible. We will present a generic observability reference architecture to illustrate this idea:

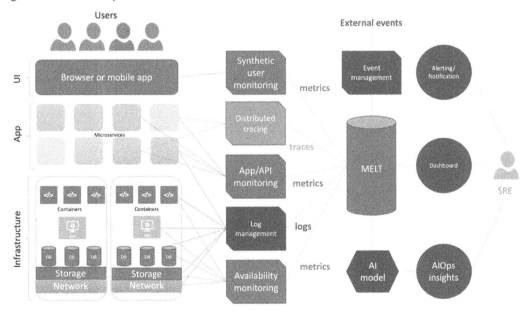

Figure 4.2 – An agnostic observability architecture

On the left side, this reference architecture displays the monitored systems, with **Users**, **App**, and **Infrastructure**. This solution has a variety of IT resources and application components, showing how diverse it can be. In the middle, we have the monitoring technologies that we have described in this chapter, and besides it, we have the **MELT** data lake, with **Event management** and a given **AI model**. On the right side, you have the visualization layer, the **Alerting/Notification** system, and **AIOps insights**. All of these are consumed and utilized by SREs.

We want to explain the effectiveness of observability next.

Observability effectiveness

To put it simply, the effectiveness of the observability principle can be given by the MTTD. The faster you detect an anomaly or disruptive event, the better. There are a couple of other aspects to include in this indicator.

Blind spots

Your monitoring system may have blind spots. Blind spots are any services, infrastructure elements, or application components that are not monitored. This also includes logs and traces that are not being captured by the monitoring tools.

SREs need to review the application diagrams, architectures, and designs to identify monitoring targets. They should also define SLIs and monitoring metrics to the same end.

A governance model must be in place to ensure changes to the system lead to changes to the monitoring configuration. Otherwise, new components or dependencies will create new blind spots. The worst way to identify a blind spot is through a call from the user to the help desk complaining about service slowness or disruptions. In these cases, you need to raise an item in the SRE backlog for immediate investigation after the incident has been resolved. Make sure the postmortems check unseen areas.

Monitoring the monitoring

What happens when the monitoring system goes down? Although it may be tempting to solve all problems by killing the messenger, most likely, this scenario will only aggravate these problems.

Monitoring the monitoring is a possible solution. We need to know when observability is in the dark and fix it quickly. Many monitoring tools also are capable of isolated self-monitoring. Nonetheless, this self-monitoring leads us to another puzzle: who monitors the monitoring that monitors the monitoring?

A much better solution is to create a robust monitoring platform that has some redundancy and self-healing capabilities. Of course, a balance between cost and reliability is necessary.

Observability's main goal is to make a system observable. We learned how it does so by applying a good mix of monitoring types, tools, and data. You are now going to dive into a practice lab next to consolidate the learning from this chapter.

In practice – applying what you have learned

To help you distill the lessons learned in this chapter, we will present you with an observability simulation lab as the last topic. This lab is based on the **Prometheus** project, an open source software suite curated by the **Cloud Native Computing Foundation** (**CNCF**), which is part of the **Linux Foundation**. There is plenty of monitoring tooling on the market that offers a developer license or community edition; we have no specific reason to use Prometheus in this lab other than its key features.

You will need this prerequisite knowledge to completely appreciate this lab:

- Familiarity with **K8s**
- Basic knowledge of the **Node.js** (JavaScript) programming language
- Good understanding of **Yet Another Markup Language** (**YAML**)

> **Important note**
> If you don't have the prerequisite knowledge, don't feel intimidated. SREs are continuous learners! Now, you know some things you need to learn based on those requirements. Please check a learning platform such as Udemy for courses that can help you.

We will divide this practice into three sections:

- Lab architecture
- Lab contents
- Lab instructions

Let's check how we have designed this lab first.

Lab architecture

We developed a simple monitoring platform and dummy application to experiment with observability in a practical way. A Prometheus-based monitoring solution is depicted in the following diagram.

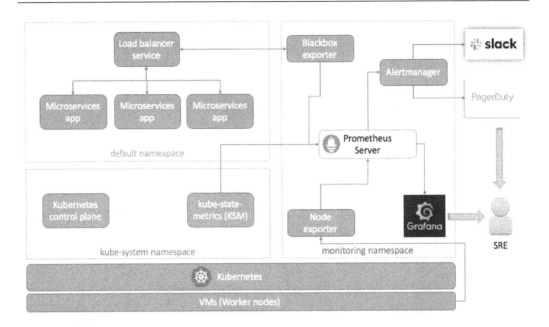

Figure 4.3 – Observability simulation lab architecture

The entire lab environment runs on a K8s cluster, except for the notification systems, Slack and PagerDuty (see *Figure 4.3*). The following components are deployed to the cluster:

- **Prometheus server**: An open source system monitoring and alerting platform originally developed by SoundCloud.

- **Alertmanager**: This handles alerts sent by the Prometheus server by deduplicating, grouping, and routing them to a notification system such as Slack or PagerDuty.

- **Black-box exporter**: This exports metrics to the Prometheus server by probing endpoints using HTTP or HTTPS.

- **Node exporter**: This exports hardware and OS metrics exposed by *nix (Unix-like) kernels to the Prometheus server.

- **kube-state-metrics (KSM)**: A K8s add-on that listens to the API server and generates metrics about the state of the objects.

- **Grafana**: An open source data visualization platform that creates dashboards for observability from Grafana Labs.

- **Microservices app**: A simple Node.js app based on microservice architecture.

The Prometheus server is responsible for pulling metrics from the targets. It does that directly from the K8s processes by using its internal service discovery feature. It also pulls metrics from exporters, which are specialized monitoring agents. Those metrics are stored as time-series data structures inside its **time-series database (TSDB)**. The process of retrieving metrics from targets is called **scraping** metrics.

The black-box exporter is an HTTP prober that monitors APIs, web services, and web-based apps. The node exporter is a Linux/Unix prober that extracts metrics from the operating system kernels.

The Prometheus server watches its rules to see whether any metric has fulfilled all conditions. If so, it fires an alert to Alertmanager with all the information for the metric. Alertmanager will send an alert to one or more notification systems using the pre-configured template. In this lab, Alertmanager is integrated with Slack and PagerDuty.

With Grafana, you can visualize the metrics stored in the TSDB through dashboards with counters, gauges, histograms, and a variety of other graph formats.

KSM is an add-on to K8s that expands the metrics collected inside a cluster. KSM is handled as an optional component under the K8s project.

Lab contents

You can clone the GitHub repository by entering the following command in your terminal:

```
$ git clone git@github.com:PacktPublishing/Becoming-a-Rockstar-
SRE.git
```

Within this repository, under the Chapter04 folder, there are two sub-folders: microservices and monitoring. Also, there's a quick setup procedure called observability-simulation-lab.md at the same level. This procedure details the installation of the **command-line interface (CLI)** tools.

After checking out the main procedure, go to the following directory:

```
$ cd Becoming-a-Rockstar-SRE/Chapter04/monitoring/prom-server
```

Inside the prometheus-configmap.yaml file, you will find the two configuration files for the Prometheus server: prometheus.rules and prometheus.yml. They are stored as a K8s ConfigMap object inside the cluster. The following extract shows how an alert rule is set for Prometheus:

```
    rules:
    - alert: HighPodMemory
        expr: (container_memory_usage_
bytes{namespace="default",image!="k8s.gcr.io/
pause:3.5",name!=""} / (1024*1024) > 14)
        for: 5m
```

```
        labels:
          severity: critical
        annotations:
          title: Pods memory usage
          description: "Pods with high memory utilization\n
VALUE = {{ printf \"%.2f\" $value}} MB\n LABELS = {{ $labels
}}"
          message: "Pods have consumed over 14 Mbytes -
(instance(s): {{  $labels.pod }})"
          summary: "Pods High Memory Usage - (instance(s): {{
$labels.pod }})"
```

The highlighted row is an expression that is calculated to check the metric threshold. In this example, it verifies whether any *pod* is consuming more than 14 MB of memory. This is not an alert that follows the user perspective notification trigger principle. Most likely, the app will consume more than 14 MB of memory and will continue to be responsive for the user.

A much better alert rule is as follows:

```
    - alert: AppHTTPResolveTimePercentile
      expr: (quantile_over_time(0.90,probe_http_duration_
seconds{instance="http://<load-balancer-vip>:60000/
fortune",phase="resolve"}[28d]) > 0.5)
      for: 5m
      labels:
        severity: critical
```

In the second example, it will calculate the percentile of the HTTP resolve time over 28 days, but only triggers an alert if the 90th percentile is greater than half second (> 0.5). This is a good example of an SLO-based alert.

Scrolling down in the same ConfigMap, you will find the global settings of the Prometheus server and where the Alertmanager pod is in the cluster. Check the following extract to see the global settings:

```
    global:
      scrape_interval: 10s
      evaluation_interval: 10s
    rule_files:
      - /etc/prometheus/prometheus.rules
    alerting:
      alertmanagers:
```

```
    - scheme: http
    static_configs:
    - targets:
        - "alertmanager-service.monitoring.svc:9093"
```

Within this same directory, `Chapter04/monitoring/prom-server`, you can find the other K8s manifest files to deploy the Prometheus server, including `Deployment`, `Service` (`NodePort`), `Ingress`, `PersistentVolumeClaim`, and `StorageClass`. Similar manifest files are available for the exporters, Grafana, and KSM.

Lab instructions

The first thing you need to run this lab is a K8s cluster. You can use any cloud provider, such as **Amazon Web Services** (**AWS**), Microsoft Azure, or GCP. We recommend using a free trial account on GCP. You can create one following this documentation: `https://cloud.google.com/free`.

You can create a **Google Kubernetes Engine** (**GKE**) cluster through the console or by issuing the following command:

```
gcloud container clusters create cluster-1 --no-enable-
autoupgrade --enable-service-externalips --enable-kubernetes-
alpha --region=<your_closest_region> --cluster-version=1.24.9-
gke.3200 --monitoring=NONE
```

The recommended GKE cluster configuration is as follows:

- **GKE mode**: Standard with a static K8s version
- **Location type**: Zonal
- **Release channel**: None
- **K8s version**: `1.24.x`
- **Number of nodes**: Three
- **Machine type**: `e2-standard-2`
- **Image type**: `cos_containerd`

Then, you need to configure your `kubectl` environment by using this command:

```
gcloud container clusters get-credentials cluster-1 --zone
<your_closest_zone> --project <your_project_id>
```

After that, you need to deploy the application:

```
$ cd Becoming-a-Rockstar-SRE/Chapter04/microservices
$ ./deploy-app.sh
```

This will create the app deployment, service, and load balancer (Google). Wait until the load balancer has been fully created and an external IP address is assigned. You can check the load balancer status by using the following command:

```
$ kubectl get svc
NAME                    TYPE            CLUSTER-IP      EXTER-
NAL-IP      PORT(S)                 AGE
kubernetes          Clus-
terIP        10.92.0.1       <none>              443/TCP             52m
node-api-rod-lb     LoadBal-
ancer    10.92.7.94      35.247.235.223    60000:32259/TCP     75s
node-api-rod-svc    NodeP-
ort        10.92.2.199     <none>              8081:31080/TCP      75s
```

As soon as the load balancer is available, you can test the app by using the following curl command:

```
curl http://35.247.235.223:60000
Hello World from Rod
```

Now that you know the app IP address, you can change the Prometheus alert rule to this assigned external IP:

```
    - alert: AppHTTPResolveTimePercentile
      expr: (quantile_over_time(0.90,probe_http_
duration_seconds{instance="http://35.247.235.223:60000/
fortune",phase="resolve"}[28d]) > 0.5)
```

And the monitoring target as well:

```
static_configs:
        - targets:
          - https://www.google.com
          - http://35.247.235.223:60000/fortune
          - https://prometheus.io
```

Next, deploy the monitoring platform by issuing this command:

```
$ cd Becoming-a-Rockstar-SRE/Chapter04/monitoring
$ ./deploy-monitoring.sh
```

Wait for the ingress objects to become available. You can check their status with this command:

```
$ kubectl get ing -n monitoring
NAME                        CLASS    HOSTS   ADDRESS         P
ORTS    AGE
grafana-basic-in-
gress          <none>    *        34.160.192.15    80       4m46s
prometheus-basic-in-
gress     <none>    *      34.160.228.230   80       4m50s
```

After the ingress objects are ready, you can access the Prometheus server and Grafana consoles in your browser. In the example here, you can open them using the following URLs:

- Prometheus: `http://34.160.228.230`

- Grafana: `http://34.160.192.15`

When accessing Grafana, the initial username and password are `admin` and `admin` again, respectively.

To import a community dashboard (so you don't need to create one from scratch), click on the **Dashboards** icon (four boxes) from the left navigation panel, and then click on the **+ import** link at the bottom. We recommend importing via `grafana.com`; the ID is `315`. Select **Prometheus** as the dashboard metrics' data source. If everything goes smoothly, you should see the following dashboard:

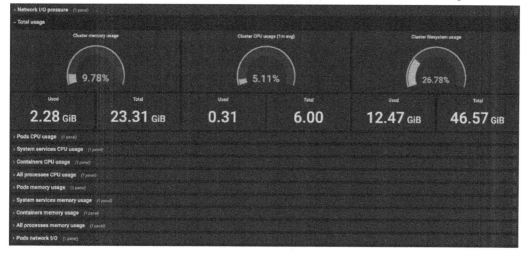

Figure 4.4 – Grafana dashboard 315

If you want to integrate Alertmanager with Slack or PagerDuty, you'll need to create a developer workspace on both. This is beyond the scope of this book. However, the Alertmanager configuration is already prepared for both; all you need is to obtain the Slack incoming webhook URL and the PagerDuty integration key.

Summary

In this chapter, we covered the intricate concept of observability and what an observable system is. There's no price to knowing how to explain what monitoring and telemetry are and how they relate to reliability. Having a good grasp on APM and how it is used to measure reliability for applications is a must-have for SREs. You heard all about recent concepts related to monitoring and event technologies. You also acquired knowledge on how alerting is done by SREs and how observability is a guiding principle at the end of the day. Finally, you consolidated the knowledge from this chapter by going through the practical simulation lab available on GitHub and understanding how to further develop it.

In the next chapter, you will learn how to approach an issue by isolating possible causes and effects, and diagnosing an anomaly when the observability platform has detected one.

Further reading

- To learn more about **Prometheus**, please check this website:

  `https://prometheus.io/docs/introduction/overview/`

- To learn more about **Grafana**, we recommend the following website:

  `https://grafana.com/grafana/`

- You can read more about observability and how it relates to DevOps in the *DevOps measurement: Monitoring and observability* page:

  `https://cloud.google.com/architecture/devops/devops-measurement-monitoring-and-observability`

5

Resolution Path – Master Troubleshooting

Every day, we see evidence of the simple fact that things break or go wrong, from the broken-down car on the side of the road to the actions we take on our mobile phones when a website is no longer working for us. And because our lives are more and more reliant on technology, understanding how to question, react to, and resolve technical issues is one of the highest-paid skill sets in the technology industry – we call it SRE.

So, how do we fix things? I've asked, *tell me what you would do if you turned ON a light switch and no light turned ON* as an interview question for years. The first answer is always to flip the switch a couple of times – and why not, our first response to many digital issues – have you restarted it? But what's next? The breaker? The bulb? Checking whether your power has been turned OFF? This seems like a lot of options, all good options, all requiring time, and all of which may not work if mice have chewed through the wiring in the walls.

The rockstar SRE looks at problems with a plan and a set of rules, accompanied by knowledge of how things work. We know to look around the room. Ah, the TV is on – so it's not the power to the house that's the issue. And look, the other two lights are working, so it's not the switch or the breaker – it must be the bulb. How did we do that so fast? Well-documented workflows and processes can help break down and resolve issues faster than ever before.

In this chapter, we are going to cover the following main topics:

- Properly defining the problem – and what to ask and not ask
- Breaking down and testing systems
- Previous and common events – checking for the simple problems
- Effective research both online and among peers
- Breaking down source code efficiently

- Logging plus code

- In practice – applying what you've learned

Properly defining the problem – and what to ask and not ask

One of the most effective steps I leverage today in troubleshooting starts even before we investigate the problem – it starts with defining an actual issue. And for good reason! How many times has someone come to us saying, "*My computer is broken*" or "*The website is down*"? Those simple descriptions can cover a gambit of different issues, including the fact that the level of severity these statements imply can be far from the true severity of the issue.

To help us translate the problem into a more actional problem statement, we'll explore multiple areas, including the source of information, variations in naming conventions, negative impacts of yes/no questions, and finally, touch on the executive summary technique, which allows us to state the question without bias or blame.

Source of information

Most of us have played or know of the children's game where you line up people and ask them to whisper something into the next person's ear, then the next whispers what they think they heard into the next ear, and so on. And in the end, it's always funny to see how that message has changed as it has traveled down a line of people. Without a doubt, this same phenomenon happens in engineering all the time. So, to get the best information, you always want to talk to the person who had the issue or someone as close to the issue as possible.

Getting information, especially about technical issues, from as close to the source as possible is not the only way to get a clearer definition of an issue – we can also ask for screenshots of desktops or messages to help us define the issue better. Talking directly to the reporter gives us the best knowledge and often, many of these reporters are familiar with the day-to-day operations we are not.

A great example of this is a report that call center agents are unable to process a payment. When it is escalated to the floor supervisor, the support desk, and then finds its way to you, the SRE. After investigating the report of "*payment processing is down*," you reach out to the original call center agent to find out that it was a single payment that worked a few minutes later when retried. These two different issues clearly require different levels of response, from all hands-on deck to a simple "well, if it happens again."

The knowledge base of the reporter

The agent from our example likely does not possess the technical knowledge of the entirety of the backend systems or even know what an API is. However, their specific knowledge of *how both normally work* is often unmatched. In fact, as a rockstar SRE, you'll want these types of people in your corner. Those who are on the front lines, using the tooling we're responsible for, can provide amazing insights into the workings and errors of the system. In addition, when resolving issues, these are the associates we want to validate that everything is working.

In addition, it's important to be aware of the technical strengths of those we work with. Everyone has specialties and areas they are weak in – even rockstar SREs. Understanding where to go and who to validate your ideas with is important. By understanding the different skill sets in question, you can feel confident in what is being said and understand when information needs to be validated by another source.

Understanding the source of the information we receive and assigning levels of trust to it will help us get a true understanding of the problem statement.

Naming conventions

Just as we need to trust the source of information, we also need to realize that not everyone speaks the same language. Honestly, how many of us have seen an entire computer case called the "*CPU*" or an LCD screen be called the "*computer*?" Indeed, even among technical people, we see load balancers called "*the F5*," or API endpoints called microservices or services – even just "*the backend*". And it's not uncommon for the wrong naming convention to be used – for example, Docker when talking about Kubernetes – or my favorite, "*we rebooted the container.*"

Like the spell-checkers of today, rockstar SREs must be able to use the context around the words to help define the meaning. This doesn't just apply to non-technical individuals; in a society where we can't even agree on what to call a can of soda/pop, this is essential.

I've also begun talking with others in their own speech – helping change their use of language but also making sure not to isolate them. For example, I might say, "*We restarted the containers, or as Mike pointed out, that was like rebooting them.*" This simple act helps others understand correct terms, as well as making them feel at ease with their choice of words.

False urgency

And finally, we need to discuss false urgency. This rather simple concept often is one of the most difficult things we deal with. From the noisy customer to the C-level executive having a minor issue – we all have dealt with situations that demand a sense of urgency out of proportion to the actual issue.

Understanding the severity of issues helps us to ensure we bring in an appropriate response. I've sat on calls that cost thousands of dollars an hour for engineers to be engaged in resolving a problem. This may be appropriate if you truly cannot accept any bill payments, but not for a single failed payment that went through minutes later.

Executive summary

We will certainly cover the executive summary with examples in *Chapter 15, Postmortem Candor – Long-Term Resolution*, as it's an important process to understand when writing postmortems – and it's simple. State the truth without blame or emotion. In short, we don't talk about who. We don't say *"Josh deployed a change and took down production"* or *"Suzie ran an ad hoc query that locked a table in the database."* Instead, we only state facts of the system – *"an ad hoc query locked a table in the database"* or *"a recent deployment caused an issue with the service."*

Why? Honestly, placing blame often either puts people in a defensive mode or makes them feel guilty inside – neither of which helps fix problems in the present. The exception to this is when you make the mistake – a rockstar SRE is always willing to admit when they break something – and sometimes, we even take the blame for others. The goal is to prevent and solve issues; discussions about people's performances are not a public subject.

We've spent a fair amount of time discussing how to identify an issue, whether understanding the source, naming conventions, or even how we restate problems as an executive summary. Now that we have a clear problem statement, we can start breaking down what's happening.

Breaking down and testing systems

In today's technology, systems have become highly complex, often reliant on multiple systems, vendor-provided services, and a plethora of data storage and management mechanisms. Checking each independent API, vendor service, and database entry can take time – a rockstar SRE knows a better way! We can segment a system into areas of functionality and test large parts of the system – **strategically testing**.

When troubleshooting a simple desktop computer or laptop, we might consider testing the hardware separate from the operating system. For larger systems, we may look at database entries or application logs to verify the expected data exists. By breaking the system apart, we can choose middle points to test and then identify which half of the system is broken. Imagine that, with one test, you could validate the functionality of the entire frontend of your website or ensure that half a dozen APIs are working. Then, you can break down the remaining broken half and continue testing halfway through the system.

Halfway-point testing may sound familiar; it's one of the standard ways of searching an ordered list. Compare the data at the halfway point, identifying whether your target is in the first or second half. Continue breaking down the list into smaller parts and testing the list's halfway point until you find your target. For a small list, this seems ineffective, but the larger the list (or system), the more efficient this method is.

Breaking down hardware versus the operating system

We shall start with a simple example, hardware versus software. Imagine you purchase another hard drive for your desktop computer, get it all hooked up, and it's not working.

So, how can we quickly tell whether it's an issue with the physical installation – the hardware – or the software – the operating system? Simple; if you are on a Windows or Linux computer, you can boot into the firmware on the main board, often referred to as the BIOS. This runs separately from the operating system and allows you to view the hardware attributes of your system, most often including hard drives.

By looking at the system's BIOS, we can tell whether the main board *sees* the new hard drive. By doing so, we can immediately identify a hardware versus an operating system issue. If the drive is not there, we can check the drive's cables, including power – if the drive does show up, we know the issue is in our operating system. Perhaps we simply forgot to use `fdisk` to define the partitions on the drive and mount the drive.

What's so remarkable about this? It's the separation; because we choose to separate the operating system from the hardware, we immediately determined the direction to take next, thus keeping us from checking every potential problem.

This concept can be even more effective applied to web APIs given the size even of simple systems. We not only save time but also solidify our standing as rockstar SREs.

Breaking down a web API

Remember our favorite custom hat website? Imagine you have a problem with generating a price; as we've discussed, we can't make sales if we can't generate a price, so troubleshooting the system needs to be done quickly. To make troubleshooting easier, the documentation includes a diagram of the parts of the system needed to generate a price (see *Figure 5.1*).

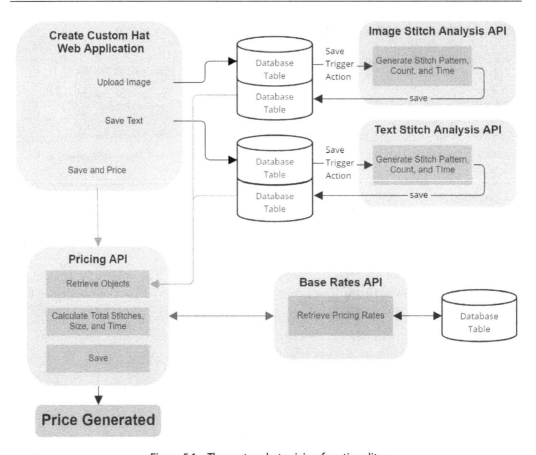

Figure 5.1 – The custom hat pricing functionality

From the diagram, we can see the system is a mixture of API microservices and database functionality:

- When data such as a text box or an image is saved to the database, the database triggers the generation of data about the item

- The Image Stitch Analysis and Text Stitch Analysis APIs both operate in the backend while the user is still working on the hat design

- When the user saves the design and requests a price, the pre-calculated information is loaded about the objects

- The base rates, such as the cost per stitch, cost by time, and overall base cost of the hat, are retrieved by an API, which retrieves the information from a database

- When all this is brought together, a price can finally be generated

By breaking the system in half, we now have a halfway point to test (*Figure 5.2*):

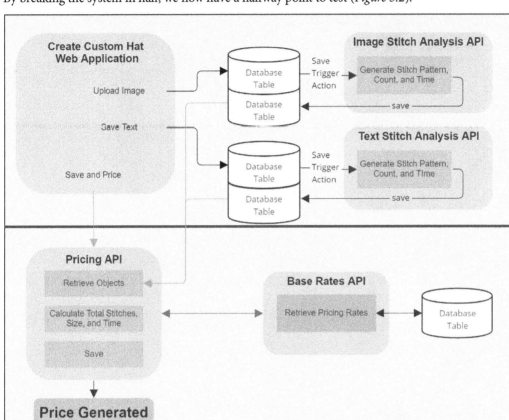

Figure 5.2 – The system broken into two halves (top and bottom)

Remember our problem? Yes, we cannot generate a price. Where do we start?

- Separate the system into two pieces; in the diagram, you can see we have split the diagram into the top and bottom parts (*Figure 5.2*)

- To ensure the top part is working, we look at the database tables to which both Analysis APIs save

- When reviewing the database tables, we do not find any data on the text

- Therefore, we know the issue is in the top half of the diagram

Once again, we halve the top half of the diagram (*Figure 5.3*):

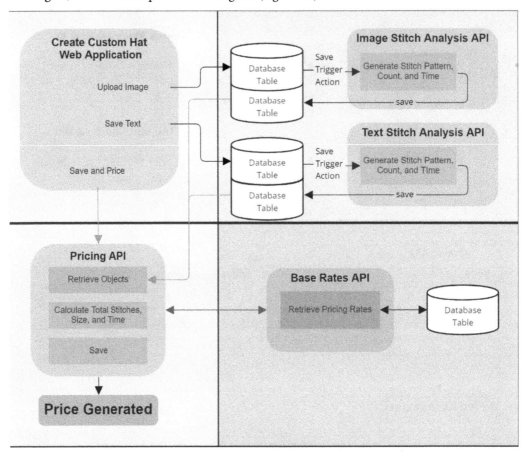

Figure 5.3 – The system halved again

Let's continue to troubleshoot:

- We look for a halfway point to test since we know the Text Stitch Analysis API is not saving data.
- We review the data in the database table to which the web application is saving the text.
- The data is present in the database table, so we know the issue is not on the left with the web application.
- This leaves us with only two things to check – the functionality of the Text Stitch Analysis API and the database trigger that calls the API.
- In reviewing the application logs, we see no entries for the API. Therefore, we identify the issue as the database not triggering the API.

Understanding the steps

The constant breakdown of this test technique allowed us to quickly identify the issue with the system in only a few steps, unlike the number of steps it would have taken us to review each part of the system.

We can review each step we took for data, logs, or metrics to ensure functionality up to that point. Remember, the idea is to quickly identify which area of the system is malfunctioning until we identify a single issue.

The problems with this method of troubleshooting

Indeed, in this example, it may have been faster to review the functionality of the two Analysis APIs one at a time. Simply starting from the beginning and looking at each step would have found the issue is a similar amount of time – in this case. Any farther into the system though and the breakdown and test method takes less time.

It should also be noted that if two issues exist in different areas of the system, while you may be tempted to start arbitrarily looking at different areas of the system, starting from the top and doing a new breakdown and test will continue to be the best time-saving method to identify issues.

If you do not have a system diagram and do not understand the flow of the system itself, this method can be difficult to carry out. Understanding the system is key to using this technique.

You may also find that when you have the same issue again and again in a system, leveraging this institutional knowledge about a system can be another effective method of troubleshooting, which we'll cover in the next in this chapter.

Previous and common events – checking for the simple problems

Ever have your car not started? What's your first thought? That's right, your battery is dead. And as you're charging your battery, you look around the dash for what you left on – starting with the headlights.

This instinct is generated by repeatedly addressing this issue, either by yourself or through the experiences of those around you. We call this institutional knowledge – and it's worth its weight in gold.

Prior Root Cause Analysis (RCA) documents

Most engineering organizations keep previous RCA documents on hand, which identify prior issues and how they were remedied. These can be useful in understanding how the system has failed in the past, although looking through them during an outage should not be part of the primary troubleshooting process.

When we look through RCA from the past, drift, or the difference between the system in its past state compared to the present state, sometimes renders information of little use, and taking similar remediation steps again can actually cause additional issues. Because of this, RCAs should be used sparingly.

Timeline analysis

When troubleshooting an issue, identifying the time at which the issue started is important. Without knowing the cause though, the timeline can be a tricky thing to use when depending on change requests, deployments, and system state changes to tie to root causes.

Issues that arise as symptoms – for example, an API timing out because a database is overwhelmed – can be a delayed reaction. The increase in load to the database could have started hours before the API started failing. As time progressed, fewer and fewer resources were available, finally choking the database to the point of no response. In this case, the root cause could have happened hours before the API started failing when a simple query was changed.

Post hoc ergo propter hoc (*after this, therefore because of this*) simply means one event caused another. It's a common logic trap – especially in troubleshooting. It needs to be clearly said – some things are just coincidences and have no causality to our problem. Let me say it again: just because something happened before the outage, it may have nothing to do with our outage.

Comparison

The last method of troubleshooting is comparison. What is different? Two different functions in an application: one works and one has issues, so what's different? Two different containers, what's different? Two different servers, what's different?

Often, we find differences by putting things side by side, although mostly these are glimpses we get of different configurations or oddities in this versus that, which are triggered at a glance.

Comparison is another logic trap. While it's useful, and can even find the issue, it's often the last resort in troubleshooting, simply due to the tremendous amount of time required given the complexities of systems today.

The best approach

So, what is the best approach? If you are the only person responding, then break it down and test, although you may want to choose to break down the system at known problem areas. Breakdowns don't always have to be half and half. If you have a response team, you can often leverage different engineers to take different paths. One takes breakdown and test, while another takes common issues, and yet another does timeline analysis.

I suggest running through examples of how a team would respond, or even doing drills, so everyone understands the troubleshooting steps available to them. Of course, the assignment of roles is also highly dependent on the skill sets of the engineers, and we'll discuss this more in *Chapter 14, Rapid Response – Outage Management Techniques.*

We've covered quite a few different techniques in troubleshooting, but sometimes, it just isn't enough. Next, we'll dive into effective search techniques – because when all else fails, there's always Google.

Effective research both online and among peers

For a seasoned SRE, there is nothing more horrifying than turning to Google during an outage. When I see a team of engineers gathered around a search engine, that means standard troubleshooting methods have been exhausted, or the issue is related to one obscure error. In these cases, if the issues aren't resolved in minutes through googling, the resolution time is often hours.

The art of the Google search

For those who haven't spent some time learning the different techniques you can use in a Google search, such as quote marks or the plus symbol, you are truly missing out. We'll cover a few basics here:

- Minus sign (-) means leave out – for example, if you were looking for the maximum speed of a Mustang horse, not the car, you could use `mustang horse speed -car`

- Plus (+) means must include – for example, to search for an error code specific to an IBM software, you could use `error 110 +IBM`

- OR between words is a "this or that" – for example, if you are looking for information on container errors, you may use `Docker OR container`

- Quotes around a group of words can tell Google to search by phrase, handy for error messages such as `"error 110: out of memory"`

Beyond these simple techniques, words matter. If you have a server running Kubernetes, it is often better to search using Kubernetes and not Docker – while very closely related, you'll find when you have an issue with Kubernetes, you'll get better results. On the flip side, if the issue is with a Dockerfile, searching with Kubernetes in the query will often lead you astray. Be on the lookout for specialized industry terms and vendor names. Searching for `AWS ACM (AWS Certificate Manager)` or simply `ACM` will yield very different results.

Skimming the content quickly and refining it

Search results must be narrowed down quickly. I no longer *read* the entire website results when I'm on the hunt for information. Often, I'm looking at example code, errors as described, target words in the description or responses, and even the rating of the answer, even using the web browser's own search function to quickly find relevant information on a page.

While the raw amount of search results you can review quickly is beneficial to a rockstar SRE, don't forget that we also look for ways to refine the search results. For example, if you look for a specific issue you are having with a load balancer such as NGINX, you may find the issue is related most to specific features, such as reverse proxying. With this information in mind, I may add the term `reverse proxy` to my original search and view those results.

Never forget your internal resources

So much tooling is built internally in organizations today, so we should always remember to look at our own internal resources. Whether simple sources such as a Git repository, README files, or more complex environments such as Confluence or SharePoint, these unique internal resources may hold valuable insight into common issues and configuration settings, or even valuable usage notes.

I have even been known to search work management software such as Jira and take a dive into Slack or Teams for information. Especially if technical help uses these channels, they can be a wealth of information.

Searching and troubleshooting discussions wouldn't be complete without diving into source code and techniques to review code, even reaching through code to drill down quickly to the actual lines of code generating the log statement – often bringing you right into the code that's having the issue.

Breaking down source code efficiently

With everything we have discussed in this chapter, we have done so with an air of speed and efficiency – after all, these techniques are often employed during downtime, when there's an impact on revenue and customers. We won't discuss how a developer would look at code – these techniques are employed when you are often given source code you've never seen before to identify issues with. You may have never even written in that language before – still, these are the rockstar-level SRE skills we employ.

Breaking down source code fast starts with having a trusted editor, whether Visual Studio Code or the full-blown Visual Studio IDE, even `emacs` or `vim` – we all have our favorites. Remember to trust your favorite editor when in the midst of troubleshooting code. I prefer Visual Studio Code or the open source variant used in GitPod or GitHub.dev.

When we search source code, while for simple items, you may get results, often searching in a Git repository will find fewer results than searching in a tool such as Visual Studio Code. The main difference often lies both in the ability to search for partial words, excellent for variables and functions made up of compound words, and searching for strings with non-letter characters, such as a dash or parentheses.

Code you've never seen

When you are troubleshooting code you've never seen, there is basic information I want to determine first. These mechanisms allow you to quickly track down different areas of code and draw you into functional areas to review.

Constants and variables are often used during startup to load configurations such as database names, external web service URLs, and even cloud assets such as object stores. For example, by identifying the constant or variable that holds the URL to an external web service, I can often search the code to quickly find the functions with that purpose.

Some editors provide the ability to view the definition of an object such as a function or provide a list of where a specific function is being used in code. I have leveraged this technique countless times to dive into complex code, drilling down through multiple function definitions to identify both the flow of a program and where similar functionality may exist in the code. This type of discovery is highly effective, but may not always provide insight into code contained in third-party libraries.

The file structure can provide immense insight into which code is where in an application. For example, anything typically labeled *helper* will often not contain the definitions for an API interface, but rather helper functions to be called inside those interfaces.

Finally, apply formatting or linting to poorly formatted code. This simply helps with readability and makes issues such as misplaced brackets easier to find.

When that fails

When all the tips and tricks fail you and you're looking for the answer to how code works, the most effective mechanism I've used in the past is to start adding comments to every line or few lines of code.

In looking at code, we often just glance, and unfortunately make assumptions. However, when we have to write comments in code, we are forced not only to actually define functionality for the content but our minds also stay a bit more idle while we type, allowing for more discovery time per line of code.

Although I rarely have to use this level of analysis on code, when I know an issue is there and I can't find it in reviewing the code, this is the tactic I use. It should also be noted that rockstar SRE pay particular attention to brackets, parentheses, and quotes – especially with nested loops.

Being able to break down and review source code you've never seen, especially in languages you've never used before, is a unique skill set, mostly learned rather than found. As our final subject in the world of troubleshooting, we'll discuss how logs and metrics intersect with code.

Logging plus code

How many of us have taken a log message and stuffed it into a Google query? It's becoming standard practice these days. However, if you take that same log statement and search through the source code for it, you'll often be brought to the tail end of the source code sending the log statement.

Not every log statement will be found in the source code as it's presented in the logs. It's not uncommon to insert variables such as indexes or IDs into logs. In fact, assigning a **Globally Unique Identifier (GUID)** to each API call and attaching it to all log statements is a common mechanism for tracing, but searching for these items can be more daunting.

Let's look at these two examples:

- `Error loading State Index 42 - Hawaii into memory`
- `Error processing customer 6ed03803-ba55-4b07-a3f0-48add6f3d8f2: Address Missing.`

Each of these log statements contains unique identifiers, the first an index and name, the second a GUID. When we see these, we must remember to simplify our search – for example, searching for `"Error loading State Index"` or `"Error processing customer"`.

Finally, when you are looking through source code for logs, remember these pointers:

- Exceptions caught by a *try-catch* can sometimes return the line number of the *try-catch* in the stack trace. If so, the line number can often be ignored.

- Stack traces are invaluable – if they're available in the error message, use them.

- When reviewing code, functions without their own error control, such as try-catches, can cause expectations to appear to come from the code calling the function.

Being able to view the code causing an error message is a highly valuable skill for an SRE, as it provides a fast drill-down into code to review where issues occur, and can help you track down issues that are not even code-related, such as an external API timing out.

In practice – applying what you've learned

In this practice lab, we will be breaking down the nightly shipping label system side of our custom hat-making business. As you will remember, every night, a separate process runs that takes the orders from the day, creates shipping labels, and sends shipping notification emails to customers. If this process fails, the system simply sends an email to notify someone of the failure, which while simple, given the non-urgent nature of this process, works well.

The following is a block diagram of the system; we'll walk through customers not getting emails one night for some of the orders as our problem statement. When we begin troubleshooting, our first step is to separate the system in half and then test the output of the processes at the point where the two halves meet:

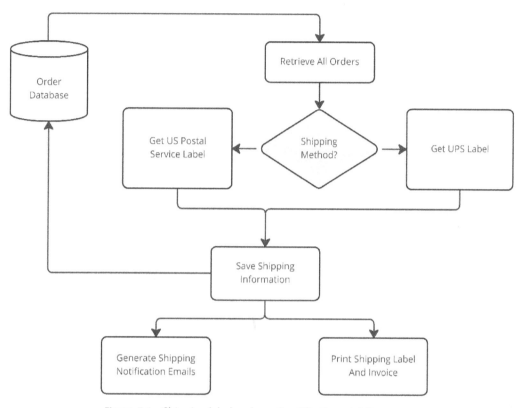

Figure 5.4 – Shipping label and email notification nightly process

The following diagram shows the system in two parts, with one part retrieving the orders and creating labels while the second part saves the shipping information, generates emails, and prints invoices and shipping labels:

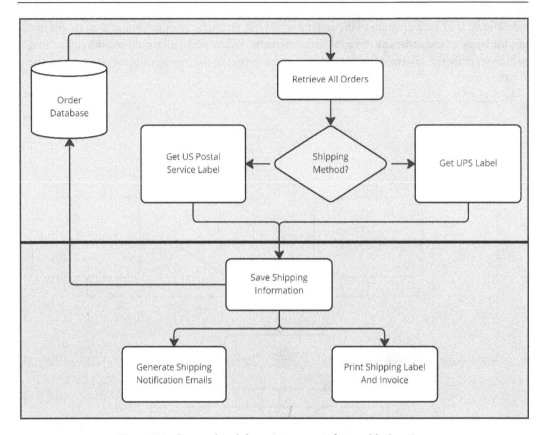

Figure 5.5 – Process breakdown in two parts for troubleshooting

When we review the diagram, we need to determine how we can see that UPS and US Postal Service labels are being generated. There are a number of ways to do this in a system, both equally useful and fast.

Reviewing the logs from the process would be an easy way to review whether tracking numbers were being generated. Depending on the complexity of the logs, this may take a long time to review.

Otherwise, we can review the tracking information generated in our shipping accounts with the shipping vendors. If you only have problems with some labels, this is a quick way to inspect whether perhaps the failure is only with one vendor.

Since this issue is with all the labels, we check via the logs that tracking numbers are being generated and find that they are. The next step is to check the database to see whether they are being saved.

If the tracking numbers were generated by the shipping vendors and the associated tracking numbers are in the logs and not in the database, we know the issue is then with saving the tracking information, so we focus on the database and the **Save Shipping Information** functionality.

As we progress through this problem, you can see we logically break the system apart, and while this is a very simple system, when you see complexity increase, it becomes more and more important to attack problems systematically. This approach reduces the number of randomly picked items to check through without thought or reason.

Remember that when we troubleshoot in this manner, understanding the flow of the system is imperative. Having these flows available to you will save time during troubleshooting – and they don't even have to be computer-generated – I've broken down a number of systems on a blank piece of paper or whiteboard.

Summary

Troubleshooting is at the heart of resolving issues and being able to perform troubleshooting with a high level of efficiency and focused methodology will drive your MTTR down. We've reviewed a number of techniques; you should be mindful of what technique you are using. When you find yourself unable to define what troubleshooting process you are employing, or worst, guessing what might fix the issue, it's time to stop, regroup, and reattack.

In the next chapter, we'll start getting deeper into the management of systems and infrastructure, an exploration into a rockstar-level SRE understanding of the foundation of the code.

Operational Framework – Managing Infrastructure and Systems

There's some confusion regarding the operational nature of **site reliability engineering**. For instance, we hear that **site reliability engineers (SREs)** exclusively work on automating *toil* or that they only manage the observability platforms that are available. Such statements cannot be true, as they defeat the very reason why we need SREs. SREs need to do operational work to handle system weaknesses, single points of failure, technical debt, performance issues, and risks. Furthermore, by getting to know them, they also fix these issues through operational work. Gene Brown, a distinguished engineer and global site reliability engineering leader at Kyndryl, once said that *"SREs need to do operational work so they can get frustrated enough by the toil they face and automate such."*

The following diagram depicts the types of work SREs do on a daily basis. They undertake operational and engineering work for applications and infrastructure. We hope this clears up what their role entails:

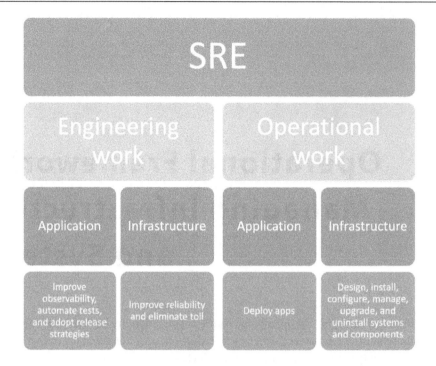

Figure 6.1 – The types of work SREs do

This chapter will address the operations facet of site reliability engineering, with a focus on the infrastructure. We want to give you a holistic view of the operational activities that SREs execute and the skillset they develop, using a disciplined systems administration approach.

In this chapter, we explain what IT systems administration is and why this is part of site reliability engineering work. We will provide information about the most adopted systems administration approaches that each SRE should know about. Since SREs are system thinkers and have a holistic view of the operational activities, we will show how they see systems and infrastructure management as a matrixial diagram. Then, we will move on to technical practices, such as **Infrastructure as Code (IaC)** and immutable infrastructures. Finally, we will close this chapter with a provisioning simulation lab to offer you practical experience.

We're going to cover the following main topics in this chapter:

- Approaching systems administration as a discipline
- Understanding IT service management
- Seeing systems administration as multiple layers and multiple towers
- Automating systems provisioning and management
- In practice – applying what you've learned

Technical requirements

Deploying a monitoring platform that conforms to the observability tenet is not difficult. Since we want accessibility over features, we have selected open source software in most cases for this monitoring system. When open source software is not available, we have indicated software that has a free offering so that anyone can gain practical experience from this chapter.

You will need the following for the lab:

- A laptop with access to the internet
- An account on a cloud service provider (we recommend **Google Cloud Platform** (**GCP**); you can create a free tier account here: `https://console.cloud.google.com/freetrial`)
- The Terraform CLI tool installed on your laptop
- The `gcloud` CLI tool installed on your laptop (if you are using GCP)
- Node.js and NPM installed on your laptop

All definition files, scripts, and application code can be found on the public GitHub repo at `https://github.com/PacktPublishing/Becoming-a-Rockstar-SRE/blob/main/Chapter06`.

Approaching systems administration as a discipline

IT systems administration encompasses a significant number of processes, operations, and tasks. Maintaining a fully operational system that has application and infrastructure components is an intense work effort.

Each component has its own level of importance in maintaining the overall reliability of the system, as well as a set of key administrative tasks to keep it healthy, available, and resilient, and ensure reasonable performance.

Also, how SREs undertake systems administration for cloud-native environments is very different from how they do the same for on-premises traditional workloads. IT resources' virtualization levels have a huge say in which administration tasks are retained by SREs and which ones are solely the responsibility of the **hyperscaler**.

We will divide systems administration tasks into seven stages, following a system component life cycle:

- Design
- Installation
- Configuration
- App deployment

- Management
- Upgrade
- Uninstallation

We will start by exploring the design of systems components.

Design

Designing a system component that is an infrastructure resource, such as a compute node, might sound strange, but each system part has requirements and specifications that must be observed to make it work with nominal performance. Besides the technical requisites, there are business demands that must be satisfied – for instance, a certain feature must be enabled for a component.

SREs must analyze these requirements, combine them, and provide a technical implementation plan for the system component. Let's check out the most common types of requirements.

Business requirements

All requirements, constraints, and policies to adhere to given by the company leadership can be interpreted as business requirements. SREs need to accommodate those requirements in an implementation plan as they are of the highest priority compared to others. A couple of business requirement examples follow:

- **Traditional on-premises system**: Hardware must be acquired from a determined vendor or list of vendors
- **Cloud-native systems**: They must enable an audit feature for certain backing (backend) services

Application requirements

Applications (**apps**) run on top of system components or consume services from them. Because of this relationship, apps may have requirements as well. SREs need to not only consider immediate requirements but also understand how app usage will grow over time if possible. A couple of app requirements follow:

- **Traditional on-premises system**: Available memory, CPU allocation, and plenty of storage space on the virtual machine of an app
- **Cloud-native systems**: Limit and request parameters for memory, CPU, and persistent volume claim allocations for application deployment on the Kubernetes engine

System requirements

A system component also comes with its minimum requirements. A component usually depends on other ones forming a stack. An **operating system** (**OS**) relies on the underlying hardware, network, and storage, while databases use OS services to function properly. It's expected that any subsystem has specifications for its dependencies. SREs must consult with a subsystem vendor to check the minimum requirements. A couple of system requirements follow:

- **Traditional on-premises system**: A minimum OS configuration and version to support a certain database system
- **Cloud-native systems**: A minimum machine type and recommended machine image for a **software-defined network** (**SDN**) virtual appliance

Security requirements

Strong security permeates all processes, including systems administration tasks. SREs check the security parameters for each subsystem. They are dictated by an organization's security policy. A couple of security requirements follow:

- **Traditional on-premises system**: Technical specifications on how to configure a middleware product to harden security and comply with the security policy
- **Cloud-native systems**: The Kubernetes cluster must follow the **Center for Internet Security** (**CIS**) Kubernetes benchmark recommendations on all worker nodes

Manageability requirements

Unfortunately, this type of requirement is often disregarded. It specifies the scripts, runbooks, tools, and automation necessary to manage a component after it is installed. SREs must ensure there will be procedures and tooling to operate this subsystem in a moment of crisis, for instance. A couple of manageability requirements follow:

- **Traditional on-premises system**: Automatic startup and shutdown scripts for the subsystem
- **Cloud-native systems**: Instructions on how to configure the `kubectl` **CLI** for a new Kubernetes cluster

Observability requirements

We need to monitor any new subsystem that is deployed in an environment. SREs need to establish how a component will be monitored and which **metrics, events, logs, and traces (MELTs)** are necessary to make this system component observable. A couple of observability requirements follow:

- **Traditional on-premises system**: Capture critical and fatal database errors from the relational database system log files and send them to the log management platform for log aggregation

- **Cloud-native systems**: Enable headers' propagation in Kubernetes sidecars and send captured headers to the distributed tracing tool for analysis

Service-level requirements

This group of requirements can be a subset of past business requirements. It discusses whether there will be any **service-level indicators (SLIs)** for a subsystem and how they are going to be measured. This is the basis for a **service-level objective (SLO)** or **service-level agreement (SLA)** later in the service performance definition process. SREs plan for SLIs before they start implementing the target system component. A couple of service level requirements follow:

- **Traditional on-premises system**: Measure the percentage of HTTP responses with 2xx, 3xx, and 4xx codes against all responses on the application load balancer in front of the web servers

- **Cloud-native systems**: Measure the percentage of HTTP responses with 2xx, 3xx, and 4xx codes against all responses on the **virtual IP (VIP)** address allocated for a multi-zone VPC

Resiliency requirements

Disasters happen, but how SREs prepare for them is key. These requirements are used to plan for mitigations and counter-measurements in case of the worst scenario. SREs consider such scenarios and prepare for them. A couple of resiliency requirements follow:

- **Traditional on-premises system**: Send certain tables' data from a database to more durable storage units, such as tapes, to keep the archives from the last five years

- **Cloud-native systems**: Use cloud storage in another zone to back up app data with 11 nines (99.99999999999%!) of durability and a long-term retention policy

Installation

Installing software is usually a simple process in the present day. It's common to see a procedure or script with the necessary commands provided by the developer to accomplish this task. You can even find a companion software installer that is based on a **graphical user interface** (**GUI**). However, we have a few software packages that still need to be compiled, built, and then installed as an OS service – for instance, **HAProxy** (https://www.haproxy.org/). SREs should be able to work with the software suite selected alongside any field of business, regardless of how complex the installation task is. Since they are pioneers in many things, they may need to create the first runbook and the first automated installation for the software installation if none are available.

> **Important note**
> We mentioned a GUI for software installers, but in most cases, we will encounter a CLI-based installation. One good reason for having a CLI-based installation over a GUI-based installation for administrative tasks is the ability to automate the former.

Installing software can be summarized as a set of steps to collect the behavior options from the user and configure the underlying resources and services consumed by it. Also, besides preparing the environment where the software will run, the installation task creates the configuration parameters for the software to run smoothly in this environment, by checking which other subsystems are present.

There are many ways of completing a software installation, and each software vendor may have a particular procedure for it. However, it should be considered as a repetitive manual task that has no enduring value, thus **toil**. SREs should always look for a method to automate this task, even with GUI-based installers.

Silent install

You may remember a time before fully automated installers were prevalent and commonplace, when we had to install **commercial off-the-shelf** (**COTS**) software manually. This was a time-consuming, interactive effort where many command lines and arguments had to be passed carefully to avoid mistakes. A mistake usually meant restarting the whole process.

The first practice that alleviated this pressure and simplified the software installation task was the silent install mode. Simply put, all the software options are placed in an installation configuration file or passed as arguments. Then, the software installer picks up those entries from this file instead of asking the SRE in the GUI. This allowed the automation of the installation task for GUI-based installers that had previously relied on user interaction.

Here is an example of the PostgreSQL database silent installation on a Windows server:

```
> postgresql-14.5-x-windows-x64.exe --mode unattended
--superpassword <DB_SU_PASS> --servicepassword <SYSTEM_PASS>
```

Provisioning

When we talk about installing IT resources such as servers (compute) and networks, we use the word **provisioning**. The reason for this is due to the fact that these infrastructure subsystems combine hardware, software, and configuration. Therefore, installing them is much more complex.

Provisioning a server used to take many days. It started with acquiring and assembling the hardware, then installing the OS, and finally, configuring and fine-tuning the server for the application that would run on top. Since the virtualization of compute resources, provisioning is about spinning off new **VMs** or creating new **Kubernetes** clusters. SREs need to have some knowledge of hypervisors and Kubernetes if they operate in a virtualized environment.

Provisioning a network used to be as troublesome as provisioning a server. We had to acquire network devices, including routers, gateways, and switches. We then placed them all together, cabled all the devices in a mesh, and then configured the network protocols to create a physical network land zone with segments (sub-networks). Everything has changed with the advent of cloud computing, where you can create virtual networks, connections, segments, and tunnels. It's possible to create entire **virtual private clouds** (**VPCs**), but SREs must have practical experience with the cloud service provider of their choice.

Many IT resources, such as databases and middleware, are offered as cloud services by a **hyperscaler**. Provisioning a database or middleware can happen by creating and configuring a cloud service. For instance, you can create a relational database service inside a cloud tenant for an app in various flavors. Some examples follow:

- **Amazon Web Services** (**AWS**): Amazon Aurora (based on MySQL), Amazon Aurora (based on PostgreSQL), MySQL, MariaDB, PostgreSQL, Oracle Database, and SQL Server

- **Google Cloud Platform** (**GCP**): Cloud SQL (MySQL, PostgreSQL, or SQL Server), Cloud Spanner, AlloyDB (based on PostgreSQL), Oracle Database, and BigQuery

- **Microsoft Azure**: Azure SQL Database, Azure databases (MySQL, MariaDB, or PostgreSQL), and SQL Server

If SREs are responsible for **data center infrastructure** (**DCI**), then they may need to understand other technologies developed for DCIs such as **OpenStack** (https://www.openstack.org/software/).

Activation

Activation is the final step when installing or provisioning IT resources. This is more a good practice than a systems administration task per se. SREs need to check the health of the installed software or provisioned server or network. You can think of this task as a pre-flight check. We recommend, at a minimum level, that the following checks happen in this task regarding the installed/provisioned component:

- Whether it is fully operational
- Whether it has been added to the inventory
- Whether it is being monitored
- Whether it has an SLI(s) defined
- Whether it has runbooks published

Configuration

Configuring software, a service, or an IT resource implies changing its characteristics and how it functions. This is a recurrent administrative task that ensures application and infrastructure components are adjusted for changing business requirements. Each piece of software, device, or appliance will have its own proprietary configuration fields and a way to store configuration information. Unfortunately, we are far from having a universal configuration framework; **JavaScript Object Notation (JSON)**, **Yet Another Markup Language (YAML)**, or a variant of both have been adopted by many software vendors.

Another key aspect of configuration is scalability. Handling a few instances of software with individual configuration files involves some effort, but imagine if you have hundreds of instances and you need to change a single configuration field for most of them. That's where automation tools come into the picture. They are vital for managing configurations when the number of components exceeds human capacity. SREs work with these tools on a daily basis, which include Ansible, Helm, Chef, Puppet, and Terraform.

Let's see a couple of other good practices of configuration that SREs should know about.

Default templates

Configuration templates can be adopted for the initial specifications of a subsystem or software. You can create a template for certain scenarios, such as the default OS parameters for the VMs in the Asia-Pacific region or the cloud database service type, and configuration for finance apps.

The challenge of using configuration templates is making sure they are being used at all. If each SRE or systems administrator is creating their own, the economy of scale is defeated. Of course, if there are custom requirements, then templates cannot be used.

Some cloud platform providers allow you to use templates for many components and services for an organization's tenant. If there's no such facility in place, you can insert **GitHub** or **GitLab** versioning control and collaboration features for templating.

Security policies

The second-best action to strengthen system security after fixing a vulnerable software version is to have a hardened configuration. Configurations that have set all parameters to favor security will make software or devices less susceptible to attacks from hackers. They can decrease a lot of the attack surface by minimizing the exposed items.

Configuration specifications are dictated by security policies and may be included in the default templates. They harden the application or infrastructure configuration. SREs must know what those specifications are and follow them through; moreover, they can also assist in defining these specifications.

Configuration drift

Many times, we change an app or infrastructure subsystem configuration to improve its performance. Other times, we do that to make a component configuration more suitable for the app requirements. Apart from the valid reasons for modifying any of the settings, we may be causing the configuration to deviate from its intended original design, moving the configuration parameters to an undesired state – for instance, making the app more susceptible to cyber attacks. This is called **configuration drift**, and SREs should take care of it by developing scanners (automation scripts) for this detection.

Config as code

Config (or configuration) as code (**CaC**) is the practice of separating an app's code base from its settings. The application config has its own life cycle and controlling processes, such as versioning, when CaC is practiced. This way, the config is handled as a software component with its isolated code repository and access control mechanism. This is especially interesting when an automated **continuous integration/continuous deployment** (**CI/CD**) pipeline is in place. Adopting CaC ensures the app config will be consistent throughout multiple application deployments. In addition, it allows for a more secure **access control list** (**ACL**) approach to configuration items, such as credentials or passwords of systems used by the app. CaC also resolves **configuration drift**.

App deployment

App deployment is the task of installing, configuring, updating, and enabling one or more applications to make a software system available for a user. Application, or app, refers to software developed for handling business requests that consume IT resources and rely on an infrastructure to function. For instance, **enterprise resource planning** (**ERP**) and **customer relationship management** (**CRM**) software are examples of applications. Another example is **Java Enterprise Edition** (**JEE**)-based applications that run on top of a middleware type called an application server.

With the virtualization of IT resources and growing cloud adoption, app deployment is becoming a much less complex task. In this scenario, a deployment process continues to install an app's business logic code in a cloud service, such as a serverless JavaScript environment. App deployment can just involve creating new **pods** in a Kubernetes-based cluster with the latest container image released by the development team.

SREs should be able to streamline the app deployment procedure into a comprehensive runbook as a first step. Later, they need to script this runbook to automate the deployment procedure using the adopted automation technology.

DevOps model

When a DevOps model is implemented in an enterprise, app pipelines are established to move committed code chunks into built and unit-tested code through a pipeline. After that, the unit-tested code is deployed to a pre-production environment for integration or regression tests through the same pipeline. If the code is approved, then it's deployed into production and becomes available to users as an integral part of the app.

Those code pipelines are also known as CI/CD pipelines or delivery pipelines. SREs need to work together with DevOps engineers to bring the observability principle to these pipelines as well. They need to make sure the app deployment is done automatically without any human intervention; otherwise, it defeats the purpose of the CI/CD pipelines.

Operational readiness review

What's remarkable about this task is the fact it became a battleground between development and operations. Too often, after an app is deployed and enabled in a production environment, it becomes the sole responsibility of the operations team. This is mainly caused by development and operations teams having conflicting goals instead of shared responsibilities. SREs are peacemakers in this war and can't take any side. To this end, they adopted the principle of having an **operational readiness review (ORR)** in place.

Management

Most operational work sits in the management domain. Keeping a system running, available, and performing accordingly 24x7 is the main goal of any operations team. It has reactive and proactive administration tasks. Reactive tasks concentrate on resolving disruptions or degradations to one or more parts, while proactive tasks focus on improving an infrastructure or an app's health, robustness, or security.

Issues

We couldn't talk about issues without citing **Murphy's Law**, which is the adage that *"anything that can go wrong will go wrong."* Another way to look at this is by firmly believing that there are no perfect systems. Issues and problems will happen, for sure, and SREs need to know how to handle them.

With a good monitoring solution that follows the observability tenet, it's most likely that anomalies can be detected before they impact a system user directly. Nevertheless, system malfunctions such as disruptions and outages can happen, and they cause the most severe shock to users. The second type is performance degradations, such as slow responses, meaningful network latencies, resource saturation, and high error rates.

SREs develop **systems thinking** and troubleshooting skills to handle issues that affect users. They need to have a complete visualization of app and infrastructure topologies, and how they are connected. Also, SREs should be aware of any recent changes to any subsystems and services. At first, they will implement a workaround and actions that will either alleviate user symptoms or restore a system to its full nominal performance. Later, they will do a more meticulous investigation of the causes of the problem, and after they have identified those causes and their factors, they will propose changes to improve the system's reliability, considering the events and how the operations team responded to them.

Health checking

Whether it's because we need to know whether a certain application or subsystem is healthy or due to a request from somebody else, we recurrently run health checks on systems. This administration task has two types; the first is called an operational health check, which is focused on the availability, functionality, and responsiveness of a system. The second type is the security health check, where we test the security parameters and settings determined by a security policy.

Operational health checks

Although most health checks should be done by a monitoring platform, there are more complex and intricate checks that are not suitable for high-frequency monitoring types. Imagine that you need to execute dozens of steps to determine whether a system is fully operational and that takes 5–10 minutes for each cycle. You won't be able to perform this heavy check using collected metrics in the monitoring platform. For such scenarios, SREs should be able to trigger the health check manually.

Sometimes, the health check runbook is not automatable for numerous reasons, such as the automation platform having limitations with GUI-based tools. Despite this, the SRE team should work to document all operational health checks and eliminate any *toil* associated with the individual steps of this task.

Security health checks

A security health check focalizes the security settings for applications, systems, and subsystems. It tackles the configuration drift that we discussed earlier. The indispensable facet of this task is comparing the current configuration with a baseline or template, or a set of expected values. This is a combinatory problem, as each IT resource and version will have a unique bucket of recommended parameter values. Keeping such models or templates may be an unending effort as new software components and versions come into the picture continuously. With that in mind, it makes sense that a community of security specialists curates content based on security specifications for a variety of products.

We recommend that SREs adopt the benchmarks from the CIS organization if nothing else is in place. They publish technical security specifications for many platforms and software products. As a collateral result, community developers have automated some of the CIS benchmarks, making them even more interesting as there's no need for creating such automations as they're already available.

Runbook documentation

SRE teams are responsible for documenting principal standard operating procedures as runbooks. This documenting effort may seem like a less important job, but it's not. Just imagine the reverse situation where there are no published and shared procedures. Either you will find challenges when handling an alert or you won't be able to understand what should be automated or automatic. The first step on the automation road is documentation. Runbooks are built in a way that makes it simple to extract the algorithms from within them and develop code for automation platforms.

In *Chapter 9, Valued Automation – Toil Discovery and Elimination*, we will cover a few ways to develop automation from standard operating procedures.

Capacity planning

This task is kind of ambiguous. It's proactive and engineering type if you look at the outcomes that are recommendations of the IT resources required for the app. On the other hand, with the cloud elasticity feature, this task becomes operational in nature but still relevant to SREs.

> **Important note**
> Enabling auto-scaling in a cloud environment doesn't defeat the need for capacity planning. One reason is the fact that auto-scaling is useful to handle temporary peaks of consumption, but it doesn't tell you about an app's future needs. Another reason is that auto-scaling can be very expensive compared to reserving machines in advance, based on growth projections.

In essence, SRE teams analyze the application consumption levels of a variety of resources, including compute, network, storage, middleware, and databases. By measuring these utilization rates over time, they can uncover the patterns and trends of resource usage. This visual data allows SREs to plan for capacity growth or reduction.

As a good practice, capacity analysis should be part of an observability strategy. Usually, we have monitoring dashboards dedicated to that purpose. Also, most cloud providers already have a built-in feature to run a capacity planning report.

Patch management

Patching software involves replacing parts of its code that are considered defective. Code bugs and technical debts lead to unexpected states or behaviors. They can even cause security vulnerabilities that expose companies to attacks. This is a sophisticated management task, as it requires pre-patching and post-patching steps.

Pre-patching work

Since components and subsystems are connected to each other to form a system, SREs should be able to identify feedback loops and causal relationships so that they can determine which parts should be stopped for a patching operation to avoid collateral damage. Patching is a coordinated effort because of that interconnectedness of the parts. Also, a specific stopping order for the components must be observed due to their loops and relationships. For instance, you need to stop the database before stopping the app that is using it.

Modifying app or software code is risky for databases and data stores in general, as it makes changes to how data is saved or retrieved. Such changes can provoke data corruption or loss if they were not well tested. Thus, creating a saving point for the elements that will be patched and affected by the patch is a good practice. This is to ensure that a rollback operation (reverting to the previous state of the software) is feasible and automatable. Besides the current code and data backups, a series of checks must happen to guarantee that the conditions necessary for the patching to work are in place. Runbooks are effective in such scenarios, and their details should be provided by the software vendor or the app development team.

Post-patching work

After the software or application is patched, it's necessary to restart the subsystems that have been stopped for this operation. Again, a specific order must be followed to restart observing dependencies and interconnections. In addition, this order needs to be part of the related runbook and possibly inside an automated script.

With the system reestablished, we need to test its main features to ensure everything is working accordingly. This is most likely a good example of toil, and therefore, it should be documented as a runbook and then automated.

If the test fails, we need to restore the system to its previous state before the patching operation. Rolling back a patch is not usually a trivial task. We recommend having a dedicated operating procedure and automaton for this task only.

If the test succeeds, then we need to replace the affected components under the monitoring watch and consider the system open for business as usual.

Upgrade

A software upgrade is like patching. Sometimes, people call a software patch an update. We prefer to think that patching is applying a quick fix to code without adding new features, whereas updating is more about adding new functionalities and correcting others in a minor release.

Unfortunately, there's no consensus on how to assign version numbers to software and applications. One good pattern that is very popular among software vendors is **Semantic Versioning** (**SemVer**) (`https://semver.org/`), which follows an `X.Y.Z` pattern:

- `X` is the major version

- `Y` is the minor version

- `Z` is the patch version

For instance, on **MySQL v5.7.39**, we use the major version **5**, the minor version **7**, and the patch version **39** of the MySQL software product.

Although there are commonalities among patching, updating, and upgrading, the latter replaces the code core upon a new major release. For instance, we may say that migrating MySQL from version **5.6** (`Z`) to version **5.7** (`Z`) is an update, but shifting it from version **5** (`Y.Z`) to version **8** (`Y.Z`) is considered an upgrade. As a *rule of thumb*, upgrades change the major version, updates change the minor version, and patches change the patch version.

Upgrades are much more complex than updates. For that reason, it's recommended to create an equivalent environment with all subsystems and apps to refine the procedure of upgrading software. It's preferrable if the dataset is similar to the production environment one. Often, upgrades are hard to automate if the target software is embedded into an appliance (also known as firmware), or if there's a big gap between the current and objective versions. For such cases, vendors prescribe doing smaller upgrades or an update first. For instance, to upgrade MongoDB from version **3.6** to **4.2**, you need first to upgrade it from **3.6** to **4.0**, and then update it from **4.0** to **4.2**.

Uninstallation

This is the cleanup administrative task when an IT resource, software, application, or system is not required anymore. Sadly, it's not infrequent to see some disregard for this task. Many times, this operation is limited to just shutting down a VM that is not required anymore. Other times, it involves getting a software suite uninstalled from a system, or deleting a deployment in a Kubernetes cluster.

To keep your documentation consistent, there is a minimal set of necessary sub-tasks to address each reference of a resource after it's uninstalled from a system:

- Archive or update the runbooks that refer to the deleted component

- Update any automation scripts used for that component

- Remove the component from the inventory

- Delete the component from any architectural or topology diagrams

- Disable or delete any monitoring metrics or SLOs related to the deleted component
- Review any contract SLAs that involve the deleted component
- Notify the DevOps, operations, and SRE teams of the removal

Other actions can be added to this list, depending on a company's policies. If this list is not observed, it certainly will lead to technical debts and toil.

SREs design, install, configure, manage, upgrade, and uninstall system components or subsystems, and deploy apps to systems, through a disciplined approach. They rely on good documentation and largely use automation to execute the required tasks. Next, we will see how processes can support systems administration as a discipline.

Understanding IT service management

Information technology service management (**ITSM**) covers IT operations and tasks as services delivered to an organization that benefits from them. This is an interesting approach to standardizing how systems administration is accomplished by distinct enterprises from diverse industries. SREs must be able to do IT management independently of an implemented ITSM framework. We will look at the most adopted framework and a variation of the ITSM model in the next two sections:

- Information Technology Infrastructure Library
- DevOps

Let's delve into the **Information Technology Infrastructure Library** (**ITIL**) framework next.

ITIL

The main goal of ITIL is to align IT services, provided by IT systems, with business needs. It does that by prescribing detailed practices for IT activities. ITIL version 4 was released in February 2019 to align its practices with modern work philosophies, such as Agile, DevOps, and Lean.

> **Important note**
> ITIL contains detailed processes, subprocesses, procedures, tasks, and checklists to help organizations of any type and industry to govern their IT services, activities, and assets.

We will describe the most important ITIL processes, starting with the incident response process.

Incident management

Any service disruption or quality reduction that was not planned is called an **incident**. The incident management process describes the inputs, outputs, and subprocesses for handling incidents. Basically, when a potential or actual failure happens, either a user, technical staff, or monitoring system detects it. An incident record is created to capture all information on symptoms, impacts, who's working on it, and the intermediate status. After the incident is open, it's assigned to a resolver; when the resolver receives it, they can acknowledge it. That means someone is aware of the incident and working on it. After the incident is resolved, the resolver runs tests on the restored IT service by troubleshooting and fixing the faulty components. If the service is functioning again, then the incident record is closed.

By having an incident management process implemented for ITSM, important performance indicators such as **mean time to acknowledge (MTTA)** and **mean time to repair (MTTR)** can be calculated using the incident record data.

Problem management

After a major incident or repetitive minor incidents occur, we need to understand what the causes are. This triggers another ITIL process called problem management that focuses on preventing incidents from happening. It starts with a problem record being created, usually due to a major incident occurrence or a frequently repeated incident. The technical staff apply RCA techniques to find out the root causes and corrective actions. After the corrective actions are implemented, the problem record is closed. SREs do the same, but they adopt a blameless approach where they focus on reliability improvements rather than just finding a cause.

Incident and problem management are interrelated and interdependent. One of the key outputs of the incident management process is the timeline – what happened when. This is an input to the problem management process. Tools that operate on chat platforms have the capability to automatically generate an incident timeline, which saves a lot of time and trouble when investigating the causes of an incident or a sequence of incidents.

Change management

The change management process aims to minimize risks and technical debts associated with changes. ITIL defines a change as *"the addition, modification, or removal of anything that could have an effect on IT services."* With that definition in mind, all administration tasks result in changes to an infrastructure, apps, or whole systems.

Standard changes are low-risk modifications that are pre-authorized. All they require is a change record (CR) documenting the implementation, test, and rollback plans. Usually, a script or automation creates a CR while implementing and testing a standard change. When an automaton does the work, the CR is used to update the timeline with the most significant events for audit readiness and further development.

Normal changes have different risk levels, such as major, significant, and minor. Each enterprise needs to define the semantics for each risk level and the required authorizations that each risk category will demand. A **CR** or **request for change** (**RFC**) is created to document the plans and risks. This RFC is analyzed by a **change approval board** (**CAB**) that can either approve or reject the change request.

Emergency changes are necessary to fix an incident or problem. They are urgent and, for that reason, cannot await CAB approval. They are verbally approved during an incident by the **emergency change approval board** (**ECAB**) on-call.

With cloud services and large-scale automation becoming familiar, and the SREs working to eliminate toil, more and more changes migrate to the standard class.

Monitoring and event management

This is also just known as event management. This process has a key goal for observability – ensuring that all **CIs** are monitored. A CI is an IT infrastructure component with a determined type such as hardware, a device, an appliance, software, an app, a network, a system, a location, a facility, middleware, a database, a service, or a cloud platform.

This is a pivotal IT governance process that verifies whether we are monitoring all components, whether the components are generating meaningful events, whether the informational events are filtered out, and whether the remaining events have an adequate response.

SREs are keen to have this process fully implemented as early as possible, through the adoption of tools that can monitor the events and correlate the CIs with the **configuration management database** (**CMDB**).

Service request management

Service requests are nothing more than requests directly from users. While change requests come from internal projects and the IT operations team, service requests are inquiries coming from the customer asking for information, access, or the provision of a new service. It's common to route such requests to a help desk that documents and tracks their status. After the service request ticket is created, it's assigned to a team or person that will serve it.

SREs must identify repetitive and automatable service requests and handle them as *toil*. A self-service portal is always a good idea to offer users a way to request services that are delivered by an automation platform without any manual labor from the operations team.

DevOps

DevOps is more a philosophy of working and a movement than anything else. Like site reliability engineering, it relies on a cultural shift, implementation of practices, development of core skills, and adherence to principles. DevOps has overlap with ITSM and ITIL, but it's still a divergent approach if we consider its areas of focus, changes in velocity, and process controls. It concentrates much more on people and technologies than processes. It embraces constant changes instead of unnecessary stability. Also, it empowers people instead of centralizing control over the processes.

The notable feature of DevOps is the pipeline where requirements are coded into commitments, integrated with the main code base, and then tested and deployed in a live environment with working software. Also known as a delivery pipeline, it relies on a toolchain (linked tools that create a code pipeline) to move code chunks from the development environment to the production platform.

As we mentioned before, site reliability engineering is much closer to DevOps than to ITIL, but SREs work with any IT governance model, whether it's ITSM, DevOps, or both.

Agile lean development

DevOps utilizes disciplined **Agile** methodologies such as **Scrum** and **eXtreme Programming** (**XP**) for the development of software or applications. It also incorporates **Lean** and **Toyota Production System** (**TPS**) practices, to name a few. Others include the following:

- **Just in time (JIT):** Minimizes inventory costs for product lines. This is applied to DevOps pipelines.

- **Andon system:** Alerts operators about an issue on the production line, hence the DevOps pipeline.

- **Ji Kotei Kanketsu (JKK):** A work philosophy that dictates defects should not go to the next process. It is applied to each step of the DevOps pipeline.

- **Kaizen:** A continuous improvement of product lines and DevOps pipelines.

- **Obeya:** A visual management of product lines, therefore DevOps pipelines as well.

- **One-piece flow:** Also known as continuous flow, this refers to the way products move from one step in a process to the next one. In DevOps pipeline terms, this is one feature or committed code chunk at a time.

Yet there's no operational work here; SREs need to know the development process, as they may need to contribute to code with monitoring instrumentation.

Continuous integration

This is a DevOps practice where developers merge their code modifications and additions to a shared repository regularly. This ensures that the main code base has all its parts and is constantly built (compiled or packaged with dependencies) and tested, so all developers will have access to the latest code changes.

If a merge leads to an integration test failure, the DevOps team has a certain timeframe to fix the bug or undo the merging.

DevOps engineers implement a DevOps pipeline that starts with the CI module. Although this is typically DevOps engineering work, SREs might maintain the pipeline themselves by using any of the processes or practices discussed so far.

Continuous testing

Testing uncovers problems in code before it's made available to users. A continuous testing process deploys working code to a live environment with very few errors. There are various types of tests, including unit tests, integration tests, regression tests, load tests, and user acceptance tests.

Each type of testing surfaces a different category of software bugs. For that reason, all tests are executed before reaching production readiness.

SREs aid DevOps engineers in developing user acceptance tests, as they can be accomplished by a synthetic user monitoring tool. Since SREs are well versed in automation, they can try to automate these tests to improve the reliability of code, thus system reliability.

We will discuss tests in more detail in *Chapter 12, Final Exam – Tests and Capacity Planning*.

Continuous delivery

If the tests were successful, then the code changes are deployed to a staging environment as a release or new version. The staging is also called a pre-production environment. From there, it's possible to run additional testing, such as an integration test, with data copied from the production environment to the staging area. Alternatively, after final approval, the code is pushed to the production platform from the staging. This is a natural evolution from the CI practice and the final part of the CI/continuous delivery acronym.

SREs are responsible for the final test and ORR before the code chunk is actually deployed into production. Undoubtedly, all tests and reviews should be automated.

Continuous deployment

The next practice is when everything is fully automated in the DevOps pipeline. After code is deployed to staging, it's automatically pushed to production and made available to end users for consumption. Of course, pushing changes automatically to production has its own risks.

One way to minimize the risks of automating deployments to production is to put into action a software release strategy. SREs are well versed in software release practices and assist DevOps engineers with these. A few practices are described here:

- **A/B testing**: This is especially good to understand user engagement when there is more than one option for a feature or functionality. In many cases, it's hard to determine upfront which application flow, feature dynamics, or user interface will have a more positive impact on users. In this practice, app versions A and B are deployed to production at the same time. SREs enable digital user experience monitoring to check which feature is more used.

- **Canary releases**: This is perhaps the most adopted strategy for app releases. In this approach, the new app version is offered to a small percentage of the users – let's say 10% of them. That way, if the new version has any problems that were not detected through the automated tests, only a small number of users will be impacted. Again, SREs ensure the monitoring captures the feedback for the new version.

- **Blue-green deployments**: This is the relatively easiest strategy, but not a cheaper one. For this one, we have two production environments – blue and green. The blue environment has the current app versions and receives all the traffic from users. The new app version is deployed to the green environment, and then the traffic is switched to it. The blue environment is still live with no traffic, and it goes on standby if green fails. If green is approved by the monitoring and user feedback, then green is re-labeled to blue.

Continuous feedback

Also called continuous monitoring and feedback, this approach means that continuous cycles of feedback should happen through the CI/CD pipeline and not just at the end. Monitoring systems are necessary to measure the success of each step or task in a delivery pipeline, and they need to provide actionable information to the DevOps engineers so that they can act upon it to add value to the user experience. That includes the monitoring of code development, deployment, and runs.

SREs are well versed in monitoring apps when they are already running in production. They need to partner with DevOps engineers to extend monitoring coverage to the inner layers of the CI/CD pipelines. That way, continuous feedback for an app will be in place.

In the next section, we describe how SREs see systems administration as a matrixial approach.

Seeing systems administration as multiple layers and multiple towers

SRE professionals follow a technical generalist model. By generalist, we meant to say they have deep knowledge of multiple technology domains. We saw in this chapter that they are also well versed in the processes and governance model that support IT operations and development. We will represent the site reliability engineering body of knowledge for systems and infrastructure management as a matrix.

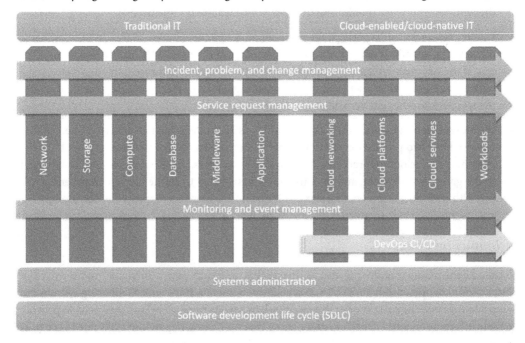

Figure 6.2 – Matrixial IT systems and an infrastructure body of knowledge

You can see in *Figure 6.2* that systems administration happens at any phase of the cloud adoption ladder, from traditional on-premises and cloud-hosted to cloud-enabled and cloud-native IT. Moreover, ITSM processes permeate not only IT models but also technology towers.

Conceptually speaking, a DevOps pipeline and its processes can be deployed to a traditional IT environment (mostly based on bare-metal and VM servers). In practical terms, the investment in creating a toolchain capable of working well in such a scenario is not justified by its return. The speed of change will be limited by the provisioning task, which takes much longer in this case. For that reason, we will add the DevOps CI/CD and embedded processes only to the cloud side of the matrix (see *Figure 6.2*).

SREs apply the systems thinking methodology to understand the emergent behaviors and outcomes from the interactions among these many components. They are passionate about learning new technologies and putting them together to foster synergy. After all, SREs were born to translate systems management complexities into a more digestible and manageable responsibility. In the next section, we talk about ways of easing some of the systems management burdens by having automated provisioning and management.

Automating systems provisioning and management

Automation is vital when we talk about systems and infrastructure administration. We don't have any other area in IT services that can easily create toil and technical debts at alarming rates. If SREs don't concentrate on making systems administration a streamlined process, soon they will not have enough people to take care of it.

There are two administrative tasks that can easily derail a whole system – provisioning and management. We will discuss the automation journey in more detail in *Chapter 9, Valued Automation – Toil Discovery and Elimination*. We will present a couple of practices that have changed the way we handle systems administration in the following sections:

- **Infrastructure as Code**
- **Immutable infrastructure**

Infrastructure as Code

The idea behind **Infrastructure as Code** (**IaC**) is very simple. What we do for apps when we code the business logic that will govern app behavior we also do for infrastructure, by instructing an orchestrator tool on how to build and configure its components. Instead of program files, we call the files with infrastructure building instructions definition or manifest files. There are plenty of software vendors that have developed IaC orchestrators, each of them having a distinct manifest language.

Declarative versus imperative

IaC is considered declarative if the final state of the target configuration is described without the details on how to get there. On the other hand, an imperative approach gives detailed instructions on how to modify an infrastructure element so that it gets to the desired state. Most tooling works with both types. For instance, on Kubernetes, we can use a declarative approach when we define a deployment inside a manifest file and create the deployment from it with the following command:

```
kubectl apply -f webapp-deployment.yaml
```

We say we are using an imperative approach if we inform Kubernetes how we created the deployment:

```
kubectl create deployment webapp --image=nginx --replicas=3
```

Pull versus push

One key part of IaC is the intended configuration of the infrastructure component. There are two ways to transfer a configuration to the recently created component – let's say a VM-based server. A more distributed way is where the configuration is pulled by the new server from a repository, while a more controlled manner is where an orchestrator pushes the configuration to a new server.

Helm

Helm (`https://helm.sh`) is a package manager for Kubernetes, and since Kubernetes already exercises IaC everywhere, it is a good example of an orchestrator. The definition files are named Helm charts, and through them, you can define, install, and upgrade IaC.

As an example, to install the Prometheus server on a K8s cluster, after you have Helm properly installed, you need to issue the following commands:

```
$ helm repo add prometheus-community https://prometheus-
community.github.io/helm-charts
$ helm repo update
$ helm install my-prometheus prometheus-community/prometheus
--version 15.13.0
```

This Helm chart will deploy the Prometheus server, Alertmanager, Pushgateway, and Node Exporter to a cluster with a standard configuration. SREs can use Helm to deploy other types of IT resources such as databases to a K8s cluster. There are 9,000+ Helm charts on Artifact Hub (`https://artifacthub.io/`) alone.

Terraform

Terraform (`https://learn.hashicorp.com/terraform`) is an IaC tool from HashiCorp®. It automates provisioning for the most common cloud providers, and it's not limited to Kubernetes engines. With it, you can create, change, or destroy any cloud infrastructure or specific service. Terraform calls its infrastructure definitions files Terraform configuration files. They have the `.tf` extension and either HCL or JSON content.

For instance, you can create and destroy a VPC network on a GCP project easily. First, you describe the provider that Terraform will invoke to connect to the hyperscaler. Providers are plugins that are maintained separately from the main Terraform code:

```
terraform {
  required_providers {
    google = {
      source = "hashicorp/google"
      version = "3.5.0"
```

```
      }
    }
  }
  provider "google" {
    credentials = file("service-account-key.json")
    project = "observability-simulation-lab"
    region  = "us-central1"
    zone    = "us-central1-c"
  }
```

Besides the project and service account key, you need to tell Terraform the GCP region and zone.

The next step is to describe the infrastructure or service, called a resource, that you want Terraform to create (or destroy):

```
  resource "google_compute_network" "vpc_network" {
    name = "rockstart-network"
  }
```

After the Terraform configuration file with all the preceding blocks and the service account key are in place, you can go ahead and use the following commands to initialize, format, verify, and create the described resource, in this case a VPC network:

```
$ terraform init
$ terraform fmt
$ terraform validate
$ terraform apply
```

This will create a VPC network on the specified region and zone. To destroy it, just issue this command:

```
$ terraform destroy
```

In a traditional on-premises IT infrastructure, creating a simple network zone like the one we created with Terraform would take days. SREs must know about the IaC available in their environment; if none is deployed, the Terraform tool is an excellent choice to deploy it.

Cloud management API

Although you can provision and manage infrastructure, components, and services via a variety of automation platforms and IaC orchestrators, you may encounter a situation where you need more granularity, security, or control for cloud platforms. In such cases, where you need to adjust parameters or execute tasks that are available through tooling, you need to develop your own code. All *hyperscalers* offer SDKs and direct access to their management API endpoints. This is another instance where SREs must develop code.

For instance, AWS has an SDK for Node.js. You can develop a simple JavaScript program to list your EC2 instances:

```javascript
// Load the AWS SDK for Node.js
const AWS = require('aws-sdk');
AWS.config.update({region: AWSRegion});
const credentials = new AWS.SharedIniFileCredentials({profile:
AWSProfile});
AWS.config.credentials = credentials;
const ec2 = new AWS.EC2({apiVersion: AWSEC2APIVer});
ec2.describeInstances(params, function(err, data) {
    if (err) {
        console.log("Error: ", err);
    } else {
      data;
    }
});
```

This is just to exemplify how powerful such SDKs are. We will explore the cloud management API in more detail in the *In practice – applying what you've learned* section with some lab practice.

Immutable infrastructure

This is a new practice that's gaining adopters. The name may lead to some confusion, as it doesn't mean the infrastructure is in stasis and no internal state changes happen. The immutable infrastructure principle means that once the infrastructure is deployed or installed, no changes can be made to its configuration whatsoever. This is interesting from a security perspective, as we assume that if a modification to any parameters is detected, it means it was not intentional and possibly a security attack.

Immutable infrastructure works with a simple design in mind. When a new VM is instantiated with an OS, middleware, database, and application parts, no changes can be made to them without deploying a new image. We cannot simply patch the OS as we would in a mutable infrastructure environment. Patching, updating, and upgrading software are not allowed here. Also, changing the software components' configurations and parameters is prohibited.

So, how do we patch, update, or upgrade any of the software components? Or how do we modify their configurations if we have a valid business need to do so? We need to create a new VM image by testing the versions and configurations working together first. Then, we will replace the old VM with a brand-new one with the requested versions and parameters.

This approach has its pitfalls, as any important information created during the running state will be lost once the VM is destroyed. Therefore, VMs and containers have a requirement to write events and logs to an external data service or an internal database. Also, it's difficult to have immutable infrastructure without IaC and virtualization practices in place.

In practice – applying what you've learned

To practice what you have read in this chapter, we will outline a provisioning simulation lab in this section. We will play a bit with Terraform, and then we will run a simple **JavaScript** app with Google Cloud SDK to do the same.

You will need prerequisite knowledge to get the most out of this lab, such as the following:

- Familiarity with cloud computing and cloud platforms
- Basic knowledge of the **Node.js** (JavaScript) programming language

We will divide this practice into three sections:

- Lab architecture
- Lab contents
- Lab instructions

Let's check how we have designed this lab first.

Lab architecture

This lab has two parts; the first one uses the HashiCorp Terraform tool while the second part is based on Google Cloud SDK for Node.js. Let's check each lab part architecture separately, but keep in mind that both use the Google Cloud API under the hood.

Terraform

The first part of this IaC provisioning lab is based on **Terraform**. We need to have Terraform installed on our laptop and a service account key generated for the project inside the GCP account:

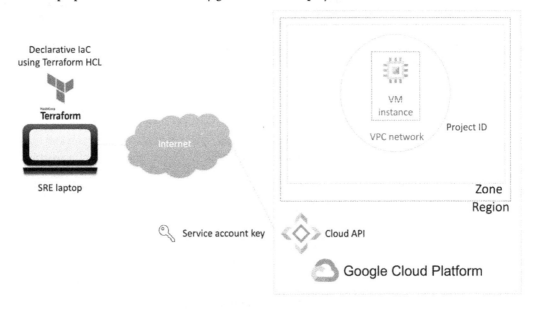

Figure 6.3 – Diagram of the first part of a lab, in Terraform

In this lab part, we are going to create a VPC network and, inside it, a VM instance.

Cloud SDK

The second part of this lab uses a simple Node.js app developed with Google Cloud SDK for the Node.js library. We need to have Node.js, Cloud SDK, and an **application default credential** (**ADC**) generated on the laptop:

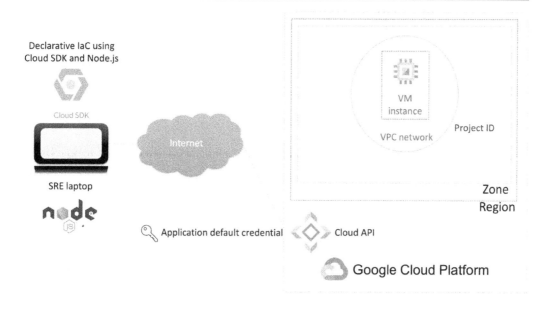

Figure 6.4 – Diagram of the second part of a lab, in Cloud SDK

In this part, we are going to create a VM instance, and then describe it. After that, we will start, stop, and delete this instance.

Lab contents

You can clone the GitHub repository by entering the following command in your terminal:

```
$ git clone git@github.com:PacktPublishing/Becoming-a-Rockstar-
SRE.git
```

Within this repository, there is a folder called Chapter06. Inside this folder, there are two sub-directories – terraform and cloud-sdk. The first part of this lab that uses Terraform is in the terraform sub-directory, while the second piece is in the cloud-sdk sub-directory.

Terraform

To access the Terraform content, go to the terraform directory:

```
$ cd Becoming-a-Rockstar-SRE/Chapter06/terraform
```

Inside this directory, you're going to find the `main.tf` file. It contains the following blocks:

- `terraform` block: This block sets up the Terraform tool by saying what the template dependencies (providers) are. It is used by the `terraform init` command:

```
terraform {
  required_providers {
    google = {
      source  = "hashicorp/google"
      version = "3.5.0"
    }
  }
}
```

The only required provider for this lab is the plugin for GCP, which is the `google` one.

- `provider` block: This second block configures the providers by telling them how to connect to the cloud platform:

```
provider "google" {
  credentials = file("project-service-account-key.json")
  project = "provisioning-simulation-lab"
  region  = "southamerica-east1"
  zone    = "southamerica-east1-a"
}
```

For the `google` provider, we need to pass the Google service account key, project ID, region, and zone.

- `resource` blocks: In these blocks, the resources are described in a declarative manner. They have descriptions of the end state of the infrastructure components:

```
resource "google_compute_network" "vpc_network" {
  name = "rockstart-network"
}

resource "google_compute_instance" "vm_instance" {
  depends_on   = [google_compute_network.vpc_network]
  name         = "rockstart-server"
  machine_type = "e2-medium"

  boot_disk {
```

```
            initialize_params {
                image = "cos-cloud/cos-arm64-101-lts"
            }
        }
        network_interface {
            network = "terraform-network"
            access_config {
            }
        }
    }
```

You see two distinct resources are declared, one VPC network and one VM instance. The VM instance has an explicit dependency on the VPC network (the depends_on statement), so the VM will only be created after the network resource is available.

Cloud SDK

For Cloud SDK, go to the following directory:

$ cd Becoming-a-Rockstar-SRE/Chapter06/cloud-sdk

Inside this directory, you are going to find the following files:

- app.js: The Node.js app code that uses Cloud SDK for the Node.js library
- package.json: The JSON file that contains the description of the Node.js app
- package-lock.json: The JSON file that contains the last state of the installed Node.js packages
- process.env-example: The shell script file that contains an example of the app configuration

app.js

Let's see the most important parts of the app. At the beginning of the file, you can find the following code block:

```
const projectId = process.env.GCP_PROJECT_ID;
const machineType = process.env.GCP_MACHINE_TYPE;
const machineImageProject = process.env.GCP_MACHINE_IMAGE_
PROJECT;
const machineImageFamily = process.env.GCP_MACHINE_IMAGE_
FAMILY;
const zone = process.env.GCP_ZONE;
const networkName = process.env.GCP_NETWORK_NAME;
```

This code block configures the parameters used by the app to connect and specify parameters to the Google API. They are gathered from environmental variables defined in the process.env file.

This statement loads the Cloud SDK library for Compute Engine:

```
const compute = require('@google-cloud/compute');
```

All functions have a similar structure by creating a Cloud SDK compute object:

```
async function createInstance(instanceName, netName) {
  const instancesClient = new compute.InstancesClient();
  ...
```

The last part is the task selector that is based on the command-line arguments:

```
switch (args[0]) {
  case 'listCustomImages':
    listCustomImages();
    break;
  case 'createInstance':
    createInstance(args[1], args[2]);
    break;
  case 'getInstance':
    getInstance(args[1]);
    break;
  case 'startInstance':
    startInstance(args[1]);
    break;
  case 'stopInstance':
    stopInstance(args[1]);
    break;
  case 'deleteInstance':
    deleteInstance(args[1]);
    break;
  default:
    console.log(`Argument ${args[0]} is not valid.`);
}
```

process.env

This file initially holds a copy of `process.env-example`. It declares the env variables used by the app:

```
export GCP_PROJECT_ID="provisioning-simulation-lab"
export GCP_MACHINE_TYPE="e2-medium"
export GCP_MACHINE_IMAGE_PROJECT="cos-cloud"
export GCP_MACHINE_IMAGE_FAMILY="cos-101-lts"
export GCP_ZONE="southamerica-east1-a"
export GCP_NETWORK_NAME="default"
```

You can modify it to accommodate the GCP project you have defined. Also, change the `GCP_MACHINE_IMAGE*` variables to use any public image you like that is available in this zone.

Lab instructions

After the lab requirements are resolved and the configuration is fixed for both parts, it's time to run the instruction to complete this lab.

Terraform

We assume that you have installed and configured Terraform here. Also, note that we have a service account key enabled for this lab:

1. Go to the lab directory:

   ```
   $ cd terraform
   ```

2. Initialize Terraform and install any dependencies:

   ```
   $ terraform init
   ```

3. Format the `main.tf` content:

   ```
   $ terraform fmt
   ```

4. Validate the Terraform configuration syntax:

   ```
   $ terraform validate
   ```

5. Provision the declared resources:

   ```
   $ terraform apply
   ```

6. Check the instantiated VM and VPC network in the GCP console (optional).

7. Delete all the declared resources:

```
$ terraform destroy
```

And we're done!

Cloud SDK

We assume here that you have followed the steps from the previous section (see the preceding *Terraform* section). Let's use Cloud SDK next:

1. Go to this lab part folder:

```
$ cd cloud-sdk
```

2. Load the env variables:

```
$ source process.env
```

3. List all custom images for the project:

```
$ node app.js listCustomImages
```

4. Create a VM instance with the name rockstar-server and a default network:

```
$ node app.js createInstance rockstar-server default
```

Check the parameters for the rockstar-server VM:

```
$ node app.js getInstance rockstar-server
```

5. Start the VM instance:

```
$ node app.js startInstance rockstar-server
```

6. Stop the VM instance:

```
$ node app.js stopInstance rockstar-server
```

7. Delete the VM instance:

```
$ node app.js deleteInstance rockstar-sever
```

And we're done!

Similar capabilities can be found in other major hyperscalers, such as AWS and Azure. SREs need to adopt the IaC approach and learn the tools that support it.

Summary

In this chapter, we learned what IT systems administration is and how to employ it as a discipline. We explained how to manage systems, services, and solutions. We rendered the multiple technology domains and showed how to navigate through them as multiple layers. We understood the concept and the usage of the IaC approach. Also, we consolidated our knowledge of this topic by going through a simulation lab available in the GitHub repository.

Now, you should know what IT systems administration is as a discipline, how SREs handle systems administration by navigating in matrixial technology domains, and how they apply IaC to provision and manage systems. You should have learned the concept of immutable infrastructure and its benefits and how to use IaC in practice.

In the next chapter, we will discuss the data science applied to observability. You will learn how to make sense of huge amounts of data.

Further readings

* To learn more about **Terraform**, check out this website: `https://learn.hashicorp.com/terraform`

* To learn more about **Google Cloud SDK**, we recommend the following website: `https://cloud.google.com/sdk/`

* You can read more about IaC in Azure's *What is infrastructure as code (IaC)?* document: `https://learn.microsoft.com/en-us/devops/deliver/what-is-infrastructure-as-code`

7
Data Consumed – Observability Data Science

Data science is a multidisciplinary field of study that contains scientific methods, processes, algorithms, and tools. SREs make decisions based on the knowledge extracted (sometimes extrapolated) from noisy datasets. They are responsible for solving problems by exercising a scientific approach within the data science domain. They don't rely on guesses or gut feelings; they make decisions based on the best data they can obtain. SREs should have a good understanding of mathematical models and statistical methods that they can apply to observability data.

We will explore and understand the most common statistical and mathematical models applied to the observability domain. At the end of the chapter, we will propose a few exercises to help consolidate this skill set in practical terms.

In this chapter, we will cover the following divisions:

- Making data-driven decisions
- Solving problems through a scientific approach
- Understanding the most common statistical methods
- Using other mathematical models in observability
- Visualizing histograms with Grafana
- In practice – applying what you've learned

Technical requirements

Deploying a data analysis platform that helps SREs to simulate data-driven decision-making is not hard. However, since we wanted accessibility over features, we selected open source software for this data analysis simulation lab. That way, anyone can gain practical knowledge from this chapter.

You will need the following for the lab:

- A laptop with access to the internet

- **Python** V3.x and **PIP** V22.x installed on your laptop

- The **Git** CLI tool installed on your laptop (or GitHub Desktop)

All files pertaining to this chapter can be found on the public GitHub at `https://github.com/PacktPublishing/Becoming-a-Rockstar-SRE/tree/main/Chapter07`.

Making data-driven decisions

Data analysis became the new gold rush not long ago, and the analysis outcomes became the gold itself. At the same time, businesses started to do **data-driven decision-making (DDDM)** by making decisions based on tangible results from solid data analysis instead of guess-estimating for business directions. SREs employ a similar approach to choosing from disparate technical courses of action, one of which will yield a better system overall and increase reliability.

SREs need to answer many questions daily, for instance, about a system's performance degrading over time. We feel tempted to probe the system a few times and come up with quick conclusions. Although this form is a fast way to answer questions, it lacks precision. This usually leads to empty efforts and unnecessary expenses.

There are trends and patterns in the data that span out over a daily timeframe. Even though SREs stay watching graphs for a whole day, they will not detect a shape that takes days to form. For that reason, they need to analyze data and make data-driven decisions.

We detail site reliability engineering DDDM as seven steps in a single process:

1. Defining the question and options

2. Determining which data to use

3. Identifying which data is already available

4. Collecting the missing data

5. Analyzing all datasets together

6. Presenting the decision as a record

7. Documenting lessons learned in the process

We start digging into the first step by defining the question and options.

Defining the question and options

The first step is to clearly define the question(s) we want to solve with data. What's the main goal in beginning this whole process? How can we translate the problem we are facing into a question? What are the possible answers or choices to this question? If there are multiple questions or matters, can we reduce them to one or prioritize one of them?

There are two imported techniques that SREs can utilize in this step. We don't think there's a better choice for stating goals and objectives than the **elevator pitch** or **objectives and key results (OKRs)**.

The elevator pitch

The idea behind this technique is rooted in an imagined situation where you want to transmit a goal to somebody important, and you only have the time of an elevator trip (less than two minutes); what would you say, and how would you say it?

There are other formats, but we love the following one:

```
AS A <role or persona>, I WANT TO <objective or requirement>,
SO THAT <reason or why>
```

We provide an example here:

```
AS A site reliability engineer, I WANT TO check whether the
application has a performance degradation over time, SO THAT I
can take preventive measures
```

In this example, you can see that it clearly communicates who wants what, what the objective is, and why we care about this goal.

Objectives and key results (OKRs)

Google adopted OKRs in the early 2000s for their company-wide quarterly and annual business goals. So we can say that OKRs and site reliability engineering walked together in the same decade, so there's no surprise if SREs make use of OKRs for their work.

We define a bold objective and the expected intermediary results, hence key results, if we move toward that objective. The format is very simple:

- **Objective**: A specific, measurable, actionable, realistic, and timely goal

- **Key results**: A list of expected intermediary results that will combine into the end objective

As an example, let's say we have the objective to check whether an app performance is lowering:

- **Objective**: Verify whether the application performance is degrading over time
- **Key result 1**: Datasets required for analysis are determined within the next day or two
- **Key result 2**: Missing datasets will be collected by the following week
- **Key result 3**: Data analysis will be completed by next week

An SRE example

After applying the elevator pitch approach, we can derive the question we want to answer and add some time precision (*last month*):

Has the application performance suffered any degradation in the last month? If yes, should we increase the IT resources used by the application or the network capacity between the app and users?

We are ready for the next step in the process.

Determining which data to use

Since we reduced the problem to a single question, we want to identify which data points and data sources we can use to help answer it. SREs look at the **metrics, events, logs, and traces** (**MELT**) datasets that may deliver on the established goal. If the question is related to a user-facing system, then they also need to check for golden signals metrics. It's important to assign a degree of importance to each dataset, for instance, high/medium/low if there are multiple sets.

An SRE example (continued)

Continuing from the previous topic example, performance degradation is mostly related to latency and saturation signals. Hence, we use those two datasets for the application with a time range of the past month.

The following are the datasets to be used:

- Application latency metrics/signals from the last month to date, high importance; data source: the monitoring platform
- Application saturation metrics/signals from the last month to date, medium importance; data source: the monitoring platform
- Application logging user waiting message from the last month to date, medium importance; data source: log management

Next, we check for existing data samples.

Identifying which data is already available

Implementing new monitoring metrics and data collectors for them has a cost, thus it's always a good idea to verify which data points are already available for consumption. Let's say some of the required data is available. SREs need to test three data aspects, at least, to ensure the existing data is useful:

- **Accessibility**: Do they have access to the data sources? The data is often available on a distinct domain or company, and access to it is not possible due to legal constraints.

- **Quality**: Does the data have acceptable accuracy and completeness? If we're analyzing one whole month, do we have data for all minutes, hours, or days within that month?

- **Aggregation**: Is the data specific to the target? Sometimes a metric is an aggregated value from multiple targets such as apps or containers. If the question is for a specific target, then this data amalgamation is not sufficiently accurate for that target.

An SRE example (continued)

Adding to the previous topic example, let's say the saturation signal for the application is present in the monitoring visualization system.

Existing data point:

- A graph with the application saturation metrics from last month to date, accessible by SREs, high quality (no missing points), and no aggregation

In the next step, we will gather the missing data, if there is any.

Collecting the missing data

Sometimes we don't have all the datasets needed to answer the initial question. Either we need to deploy a new monitoring configuration to start collecting a missing metric or look for alternative data sources. It may be a disappointment to not have all the data samples initially, but it is a reason to review the monitoring platform as a single process. SREs check for missing MELT and golden signals and correct the gaps. If it's not possible to wait to gather new data, SREs extrapolate data points from the existing datasets but never go on guessing. If there's no data to reasonably support a decision, then we consider it an experiment to gain knowledge and not a DDDM process.

An SRE example (continued)

Continuing the previous example, we determined that there's no logging for user waiting time and state instrumented in the application code. This will take a considerable amount of time, but a request is opened to the development and SRE teams. Meanwhile, we discover that some application latency data is available in the monitoring platform database; however, it's not in any form of visualization and has not been analyzed.

Collected data points:

- Application logging user waiting message cannot be collected; it will be implemented in the next iteration
- Application latency metrics/signals from last month to date collected

Next, we start to analyze the data we have.

Analyzing all datasets together

SREs can use a variety of data analysis tools such as Tableau, Python, SQL, or R to examine the data points and correlate them. There are many other tools available, such as **Software as a Service (SaaS)** or even cloud services offered by a **hyperscaler**. Also, monitoring visualization dashboards can be used to manually correlate datasets. Nevertheless, some knowledge of data science, including mathematical models and statistical methods, is necessary to achieve this step properly.

After the analysis is done, present the outcomes as insights. Insight is a deep understanding of the datasets obtained through one or more methods. For instance, a good insight would be *we observed an increase of 25% in the requests for the application during the night shift in the last 8 days*.

An SRE example (continued)

Since we have already investigated the saturation metric, we concluded that there were no significant saturation periods. Therefore, the application didn't exhaust its IT resources for long enough to affect the user. We don't see any correlation to other metrics, so we move to the latency time series data. We use Python to check the data distribution as a histogram, and determine that 10% of the time, the app latency is above 3 seconds. We check the user feedback reports and see a correlation between high latency and low scores in that report.

Insights:

- Application saturation metric displays no significative findings
- Application latency metric shows 10% of the time is above 3 seconds
- User feedback reports have lower scores that correlate to the high latency times

Presenting the decision as a record

Depending on the situation, the DDDM results are used to justify an investment or a business process related to revenue change. The more valuable the decision is, the more detailed the thought process has to be. SREs are good at communicating and articulating expensive change requests and improvement requirements. A neat way to present the results is by documenting them as insights.

One possible form to use for describing an insight is given here:

- **Issue**: What is the problem or question?

- **Context**: What is the context and background for the problem or question?

- **Motivation**: Why an analysis and/or synthesis was inevitable?

- **Arguments**: What are the results from the analyses or syntheses?

- **Decision**: What is the decision based on the arguments?

- **Consequences**: What are the consequences of this decision?

- **Next steps**: What are the next steps after a decision has been made?

This template is an adaptation of the **architecture decision record** (**ADR**) methodology, which architects largely use to document critical decisions in the solution design. Other ADR templates may be adapted for a DDDM; the principal is documenting them.

An SRE example (continued)

Finally, based on the insights from the data analysis, we document the following DDDM for our example:

- **Issue**: Has the application performance suffered any degradation in the last month? If yes, should we increase the IT resources used by the application or the network capacity between the app and users?

- **Context**: There was no formal user complaint, but the user feedback report shows a few low scores.

- **Motivation**: It's imperative that we investigate whether we have a performance issue.

- **Arguments**: The application saturation metric displays no significative findings, the application latency metric shows 10% of the time is above 3 seconds, and the user feedback report has lower scores that correlate to the high latency times.

- **Decision**: Invest in network **quality of service** (**QoS**) allocated bandwidth between the data center and the users' principal locations.

- **Consequences**: An increase in networking costs.

- **Next steps**: Determine the increased amount and start the procurement process.

Documenting the lessons learned in the process

Many insights and records are helpful for future iterations or even for other projects or apps. Documenting them in a central repository is an excellent idea. It also helps limit this type of *toil* by allowing SREs to utilize the same tools and programs.

SREs make decisions all the time and they genuinely try to use the DDDM approach for most of them. In the next section, we move on to solving problems using a scientific approach as part of the data science skill set.

Solving problems through a scientific approach

Solving problems is part of any human being's life, whether you like it or not. With the growing complexity of IT systems, a more meticulous approach to solving problems on multiple domains and layers is mandated. We want to apply a scientific method to resolve such issues. Moreover, as SREs, we use **empiricism** as the basis to prove our hypotheses and mental models that explain why a problem is happening.

The scientific method is an empirical and cyclical process, as shown in the following diagram:

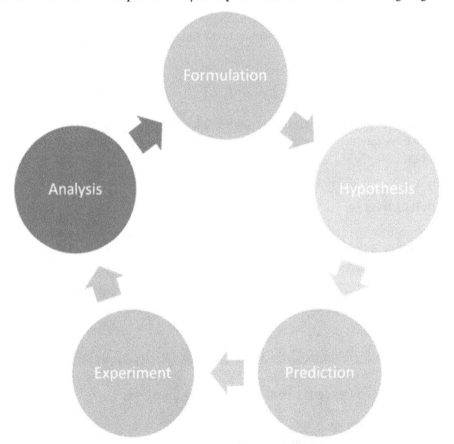

Figure 7.1 – The scientific method process

Let's understand each step in the scientific method process, starting with formulation.

Formulation

Like DDDM, we want to reduce the problem statement into a question or a few questions. This step involves evaluating evidence from any information systems that hold data from the problem domain. If there are former data analyses or documented insights for the problem, we also need to have them assessed. Based on the current evidence and previous conclusions, we formulate a question whose answer is the explanation of the problem. For instance, if the number of customer complaints increments over the previous night, a good question for this problem statement would be as follows:

- **Formulation**: Why has the number of user-initiated tickets increased in the last 10 hours and why are most of those tickets from a single region?

Hypothesis

The hypothesis is just a supposition of a possible explanation for the observed phenomenon. After examining the evidence in the last step, we need to determine a mental model or conjecture to define the forces behind the detected behavior. Of course, seasoned SREs with broad practical experience will think of better hypotheses than new SREs. Nevertheless, SREs should apply systems thinking methods to devise a model capable of inferring the noticed system comportments. As an example, say there's a user service performance issue in the system that started 10 hours ago:

- **Mental model (hypothesis)**: One or more user-facing services are having performance issues that started 10 hours ago or so. They depend on zone-specific cloud resources.

In the next topic, we will try to predict the output if this hypothesis is assumed true.

Prediction

Assuming our hypothesis is true, what would the expected logical outcomes be? Since we created a mental model to explain the problem, we need to deduce the direct consequences of having that model in place. For instance, if our hypothesis is a performance issue in one of the user-facing services within the system, we would expect to see anomalies in the four golden signals in the monitoring system for this system. This is how we validate any hypothesis by checking whether its expected consequences have happened:

- **Prediction**: There are anomalies in any of the four golden signals for the system with a problem, but they didn't go over the minimum threshold to trigger an alert. Those anomalies are isolated to a single region or zone.

Now let's discuss the experiments we can run to gather data to support a model.

Experiment

Now that our hypothesis has been minimally tested, we can run an experiment to collect data and ensure the supposition is secure. Following the example from the prior topic, let's say we want to run synthetic user scripts to measure the process times for a typical user in various services:

- **Experiment**: Run synthetic user scripts to mimic a real user and measure the process times for each transaction. Do this for different cloud zones.

Next, we inspect the data we have collected from tests and experiments.

Analysis

After we have the data from the experiments or tests, we analyze them. We can apply mathematical models or statistical methods to datasets to uncover the relationships among them. The analysis results will either confirm our hypothesis or refute it. If the insights from this data analysis don't support the premise dictated by the mental model, then we need to reformulate the problem or maybe prepare another hypothesis. We analyze the data points from the previous example and conclude that our suppositions are correct. There's a service degradation only for one of the zones. Further investigation shows a misconfiguration of a cloud service in that zone:

- **Data analysis**: The hypothesis is correct. There is service degradation in one zone due to the misconfiguration of a cloud service.

We mentioned statistical methods a few times in this section, so let's examine them in the next one.

Understanding the most common statistical methods

Statistics is a section of the applied mathematics discipline. It's used to analyze and infer conclusions from quantitative datasets. Statistical methods are largely used to understand data points and how they correlate to each other. This is exactly what SREs need to have in their skill set for the observability data.

This is a vast knowledge area that we will only scratch the surface of in this book. We want to give enough information to start using statistical methods in daily activities. We will divide the contents of this subject into four sections:

- Percentages
- Mean, average, and standard deviation
- Quantile and percentiles
- Histograms

We will do a quick tour of the statistics world next.

Percentages

Simply put, the percentage is a ratio measurement expressed as a fraction of 100. It represents a real number from 0 to 1 as a portion within 0 and 100 percent. It's perhaps the most utilized way to display gauges where the percentage indicates how far (or close) a metric is from either empty (0 percent) or full (100 percent).

SREs also employ percentages on service level indicators to determine how good or bad the service performs. Let's see this example in more detail in the next section.

Service level indicators (SLIs)

A **service level indicator (SLI)** is a measurement reported as a percentage for a service performance metric. As a convention, we say if an SLI is 0% (or 0.00), then the measured service aspect is disastrous. On the other hand, an SLI of 100% (or 1.00) means a perfect performance for the characteristic.

The equation for SLIs is as follows:

$$SLI = \left(\frac{good\ events}{valid\ events}\right) \times 100\%$$

Figure 7.2 – The SLI equation

We divide the measured good events by the total valid events. If we don't multiply it by 100%, we have a number between 0 and 1. The precision is proportional to the number of data points. If we have thousands of samples, we can use 2 decimal points after the period, such as *78.56%*.

The trick here is to define a good event. This is a combination of metric and minimal performance. Imagine that an SLI for latency in the mobile app is required. The metric is the measured latency from the observer to the app load balancer in milliseconds. The minimal performance is below 1,000 ms or less than 1 second. In that case, the SLI is the proportion of requests within 1,000 ms measured on the load balancer for the mobile app.

Next, we will talk about descriptive statistical methods.

Mean, average, and standard deviation

In this section, we will discuss descriptive statistical methods. Such methods help to describe the properties of datasets. Descriptive statistics focus on aspects such as tendency and variability, which can tell us how trustworthy the data sample is. Let's understand three of them next.

> **Important note**
> The main types of statistical analysis are descriptive and inferential. However, there are at least another five types described in the literature. It's not this book's intention to explore the entire statistical domain but to give SREs minimal knowledge on this matter.

Mean

Mean is a measurement of central tendency that is nothing more than an attempt to describe an entire dataset through a value. In this case, what's the middle of a dataset? This is a good indication of a service throughput or characteristic. If we want the system performance to be over 1,000 requests/second, but the mean of the measured requests in the time frame is around 100 requests/second, then we have work to do.

Multiple types of mean can be used depending on the data points and samples you have at your disposal. There are arithmetic, geometric, weighted, and harmonic mean types to mention some of them. Each has its own equation, advantages, and disadvantages.

Average

Also known as the arithmetic mean, the **average** is one of the first statistical methods that we learn in school. The average equation is as follows:

$$A. \, Mean(\bar{x}) = \frac{1}{n} \times \sum_{i=1}^{n} x_i$$

Figure 7.3 – The arithmetic mean equation

The values of a dataset are summed up and divided by the number of elements. For instance, let's say the error rate measures for a web app are in the following list:

```
0.054, 0.048, 0.043, 0.052, 0.049, 0.064, 0.062, 0.047, 0.045,
0.044
```

The average is achieved by adding all of the values divided by the number of elements. The total is `0.508`, and the number of elements is `10`. Therefore, the average is `0.0508` or *5.08%*. A possible interpretation for this dataset is the web app gives 5% of error codes on average.

Standard deviation

Standard deviation, denoted by the Greek symbol sigma (*s*), is the measurement of a dataset dispersion. The sigma equation is as follows:

$$Std \, Deviation(\sigma) = \sqrt{\frac{\sum_{i=1}^{n}(x_i - \bar{x})^2}{n - 1}}$$

Figure 7.4 – Standard deviation equation

We compute the dispersion by calculating the difference between each data point and the dataset mean, summing them up, dividing by the number of samples, and getting the square root. The farther those distances are from the average, the more the data is spread out. In observability, a high scattering of data points may indicate an intermittent problem that the monitoring system most likely does not capture.

From the previous section example, the mean is 0.0508, and the number of data samples is 10. We can calculate the standard deviation using an Excel sheet with the STDEV.S formula, which gives a result of 0.0073. That indicates a low dispersion of this dataset since *s* is small compared to the mean.

Quantiles and percentiles

Quantiles (and percentiles) belong to the inferential type of statistical method. We use inferential statistics to draw conclusions about the dataset's features. They also aid in determining the certainty of such inferences. It all starts by understanding that any collected data has a probabilistic distribution, which gives the probability of a value happening. In our case, this is the likelihood of a monitoring metric value occurring.

Some quantiles receive special names, such as the 4-quantiles, which are called **quartiles,** and the 100-quantiles, which are called **percentiles**. Therefore, the 2nd quartile is equivalent to the 50th percentile, while the 75th percentile is the same as the 3rd quartile.

After a dataset is treated as a distribution, and its probabilistic functions are calculated by doing a linear regression, we slice the distribution range into equal parts called quantiles. Say the data points are grades from students. We can answer questions such as which score a student must make on a test to be in the top 10% of all students. For that, we calculate the 90th percentile. If the data points are latency values measured during the last 3 days, we can compute the 99th percentile to determine the app latency for the highest 1% of the cases. This returns the value that separates the bottom 99% from the top 1% in the distribution.

We can calculate percentiles with the Python NumPy library. For instance, you can find the 65th percentile of distribution with the following example code:

```
import numpy
distribution = [13,21,21,40,42,48,55,72]
65percentile = numpy.percentile(values, 65)
print(x)
```

We will explore Python statistical libraries in the lab at the end of this chapter. Let's talk about visualizing those distributions in the next section.

Histograms

A histogram is a visual representation of numerical data distribution. It tells you the *shape* of the analyzed dataset. We use histograms to observe how quantiles (or percentiles) are distributed in the dataset. They add good value to visual analysis for datasets.

We used the Python Matplotlib library to render the histogram graph in *Figure 7.5*.

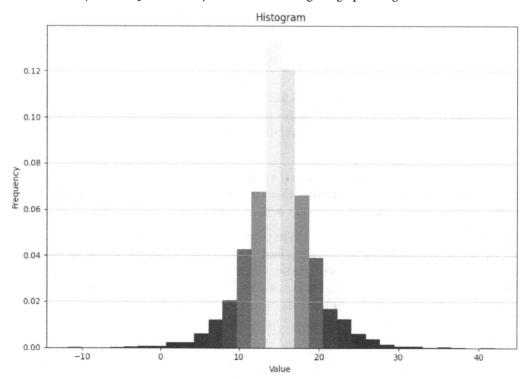

Figure 7.5 – A histogram plotted from a Laplace function distribution

We used a random number generator function that utilizes the Laplace equation to generate data samples. It creates a symmetric and unimodal distribution that's good enough for learning purposes. On this histogram, we can see that the median is around the value of 15. Most of the values are between 10 and 20, but there are occurrences of amounts around 40 and -10. We can say that more than 95% of the values are between 0 and 30; therefore, there's a 95% chance the next value will be in that range.

We detail this histogram in our lab at the end. Next, we will quickly debrief on other possible mathematical models.

Using other mathematical models in observability

SREs can go beyond traditional statistical methods by either applying computer-based simulation or artificial intelligence to extract augmented knowledge from monitoring data points. Computer-based simulation can run uncountable scenarios against a statistical model to forecast results. Machine learning employs dataset modeling to identify patterns usually hidden inside them. Let's talk about these two examples next.

The Monte Carlo simulation

This is an algorithm that simulates an event with randomly generated inputs, calculates the output based on a model, and repeats this cycle many times. The **Monte Carlo** algorithm estimates the probability of an event happening after thousands of simulations with random variables. For instance, we could create a model of an app to estimate the probability of a component taking more than 2 seconds to respond after 10,000 simulations with random values as input. We can use Python with the `matplotlib` library to create simulations with the Monte Carlo algorithm.

Machine learning

Machine learning (ML) ingests data samples to produce data models. It's capable of modeling how an app or component behaves in front of certain events. For instance, an ML model can determine the expected overall latency based on multiple inputs from distinct data sources.

SREs should know that advanced mathematical models such as computer-based simulations and ML create the basis for problem predictiveness. Next, we check how open source visualization tools such as Grafana can help with histograms.

Visualizing histograms with Grafana

Grafana is one of the most adopted open source interactive data visualization platforms. You can visualize many types of data from multiple sources using a powerful query language called **PromQL**.

It's especially good at time series data analysis, such as histograms, where we want to check for quantiles and percentiles. You can play with the Grafana sandbox environment by going to `https://grafana.com/grafana` and then clicking on the **Grafana sandbox** button.

You can check out the Grafana console's main features and data analysis demonstrations inside the sandbox environment. Click on **1 - Time series graphs** under the **Feature showcases** box. You can learn about the various types of graphical analysis, such as interpolation, soft min and max, stacking, bars, and histograms. You can access further actions by clicking on the graph title, including view, edit, explore, and inspect.

SREs should be versed in data visualization platforms to quickly make sense of vast amounts of data from the full stack monitoring platform. By transforming monitoring metrics, events, logs, and traces into graphs and graphical analysis, SREs can obtain a good arsenal for detecting system behavioral anomalies and problems.

We have explained some of the mathematics and statistics behind the visualization so you can understand how those graphs are calculated. However, in ordinary SRE daily life, we use a visualization platform to ingest the monitoring data and transform it into aggregations that humans can read and analyze on the fly.

In the next section, we will start our data analysis simulation lab to fix the mathematics from data science.

In practice – applying what you've learned

To practice what you have learned in this chapter, we have prepared a data analysis simulation lab in this section. We will run a simple Python code to analyze latency data from three different applications.

You will need prerequisite knowledge to completely appreciate this lab, such as basic notions of the **Python** programming language.

We will divide this practice into three sections:

- Lab architecture
- Lab contents
- Lab instructions

First, let's check out how we have designed this lab.

Lab architecture

This lab uses Python with two distinct libraries from the **Python Package Index** (**PyPi**): NumPy and matplotlib. You will need to have the python and pip binaries installed on a laptop with access to the internet (see *Figure 7.6*).

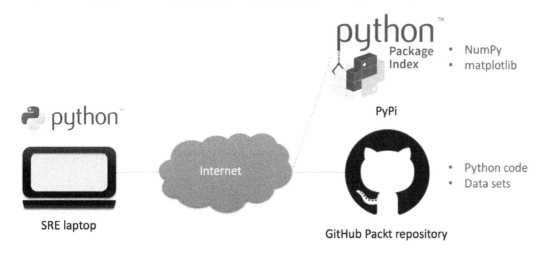

Figure 7.6 – The lab architecture for the data analysis simulation

The `pip` command pulls the packages from PyPi and installs them on the laptop. After `pip` configures the libraries, we use the `git` command to clone the repository located in GitHub that contains the Python code and datasets used in this lab.

Lab contents

You can clone the GitHub repository by entering the following command in your terminal:

```
$ git clone git@github.com:PacktPublishing/Becoming-a-Rockstar-
SRE.git
```

Within this repository, there is a folder called `Chapter07`. All the content for this chapter resides there. The first part of this lab uses the `histogram.py` source code file in the Python subdirectory. The second piece of this lab utilizes the `latency.py` and `latency.csv` files in the same subdirectory

histogram.py

To access the Python content, go to this directory:

```
$ cd Becoming-a-Rockstar-SRE/Chapter07/python
```

Inside this directory, you're going to find the `histogram.py` file. It contains the following blocks.

Dependencies block

We declare the Python libraries that this code depends on in this block:

```
import warnings
warnings.simplefilter("ignore",
category=PendingDeprecationWarning)
warnings.simplefilter("ignore", category=DeprecationWarning)
import matplotlib.pyplot as plt
import numpy as np
from matplotlib import colors
```

We are importing the two libraries mentioned. If you haven't installed them yet, you must execute the following commands before you can use them:

```
$ pip install numpy
$ pip install matplotlib
```

Dataset generation block

In this block, we have the random generation of a dataset with a specific distribution shape (Laplace equation):

```
np.random.default_rng(1234567)
dist = np.random.default_rng().laplace(loc=15, scale=3,
size=5000)
```

Declarative statistical block

We use NumPy functions to calculate the arithmetic mean, weighted mean, and standard deviation of the generated data points:

```
am = np.mean(dist)
wm = np.average(dist)
std = np.std(dist)
```

Inferential statistical methods

We use NumPy functions to calculate the arithmetic mean, weighted mean, and standard deviation of the generated data points:

```
q99 = np.quantile(dist, 0.99)
p85 = np.percentile(dist, 85)
```

Plotting block

Finally, we use the `hist` function from `matplotlib` to compute the histogram and plot it. Then the `show` function displays it:

```
N, bins, patches = plt.hist(dist, bins=30, density=True)
...
plt.show()
```

Notice that `matplotlib` uses the NumPy function to calculate the histogram.

latency.py

This Python code reads a **comma-separated values** (**CSV**) file and uses the read metric values from the CSV as a dataset for distribution. It calculates the 50th percentile of the sample and the equivalent histogram:

```
...
P = 50
percentile = np.percentile(dist, P)
print(P,"-th Percentile: ", percentile)
...
N, bins, patches = plt.hist(dist, bins=20)
```

We use the same Python functions as the `histogram.py` code except for the `dist` array, which is populated with the values from a CSV file.

latency.csv

This file contains 1,000 data samples of an app-measured latency metric. It contains five columns: `timestamp`, the `metric` name, the `configuration_item` name, metric `unit`, and metric `value`:

```
timestamp,metric,configuration_item,unit,value
1664824901,latency,app01,ms,1033
1664824961,latency,app01,ms,620
1664825021,latency,app01,ms,1339
1664825081,latency,app01,ms,614
1664825141,latency,app01,ms,1100
1664825201,latency,app01,ms,1319
1664825261,latency,app01,ms,3292
1664825321,latency,app01,ms,1161
1664825381,latency,app01,ms,1169
...
```

The timestamp is given in the **Unix epoch** format, which is the number of seconds since 00:00 on January 1st, 1970.

Lab instructions

After the lab requirements are in place, it's time to run it. The first part is to visualize a histogram. Since we are using a mockup dataset generated on execution time, all you need to do is to issue the following commands:

```
$ cd python
$ python histogram.py
```

The preceding command will render a histogram on your screen.

For the second part, we provide a dataset of measured latencies on an app component. You need to figure out the percentile of the latency over three seconds. You need to adjust the P constant to find which one brings the answer:

```
// latency.py
P = 50
percentile = np.percentile(dist, P)
```

After modifying the P constant, rerun the code to see what the latency is by using the following command:

```
$ python latency.py latency.csv
```

We hope you enjoyed this lab!

Summary

In this chapter, we discussed how SREs take and give directions based on insights and knowledge extracted from data. We outlined how they resolve complex problems by troubleshooting them with a scientific method. Together, we understood the most used statistical methods to analyze observability data to uncover patterns, trends, and anomalies.

Additionally, we acquired knowledge of other mathematical models SREs can apply to datasets. We consolidated the knowledge from this chapter by going through the practical simulation lab available on GitHub. Finally, we showed how to further develop your data science skill set.

In the next chapter, we will discuss reliable architectures and how to apply strategies to have good system designs.

Further reading

- To learn more about Python for histograms, please check out this website: `https://realpython.com/python-histograms/`

- To learn more about the Python **NumPy** library, we recommend the following website: `https://numpy.org/doc/stable`

- And for the Python `matplotlib` library, please use this link: `https://matplotlib.org/stable/tutorials/`

- You can take a course on data science on Coursera at this link: `https://www.coursera.org/browse/data-science`

- If you want to learn Grafana, this is the first tutorial you should take: `https://grafana.com/tutorials/grafana-fundamentals/`

Part 3 -
Applying Architecture
for Reliability

Developing systematic thinking and being able to apply systems-thinking methodologies is a must-have skill inside the reliability architecture design domain before everything else. First, we discuss architecture from serverless to containerization, load balancers, and how the work is done in the most reliable ways. Then, we examine reliability across multiple types of architecture and strategies, such as blue-green deployments and canary releases, and check microservices and how to make them highly available and reliable.

The following chapters will be covered in this section:

Reliable Architecture – Systems Strategy and Design

As most of you should know by now, reliability doesn't start in the app code development but in the solution architecture design. As we said at the beginning, SREs also have architectural skills. They need to review architectural diagrams, application topologies, and solution designs to spot areas where reliability is unclear.

In this chapter, we will cover the following main sections:

- Designing for reliability
- Splitting and balancing the workload
- Failing over – almost as good
- Scaling up and out – horizontal versus vertical
- In practice – applying what you've learned

In this chapter, we will define how to break up the workload among multiple workers, whether servers, containers, or serverless. Then, we will discuss the load balancer and how it enables reliability and follow up with a discussion about what is better: bigger workers or more of them. Moreover, what are the pros and cons of each? Besides that, we will look at how spreading the load across multiple zones and geographical areas is vital for planet-scale applications.

Also, this chapter will explain the reliable design and good patterns when architecting for reliability. We will help you identify how we split and balance load across different types of workers in a data center, zones, and geographical regions and the benefits and advantages of each arrangement. Additionally, we will review why failover is sometimes the only choice, and how to automate failovers when an irreversible problem is detected. We will illustrate how to scale out workloads and the types of scaling. Finally, we close this chapter with a scaling-out simulation lab to consolidate your knowledge.

Technical requirements

At the end of this chapter, you'll find an autoscaling simulation lab based on cloud infrastructure, as we want to ensure you leave this chapter with practical knowledge.

You will need the following for the lab:

- A laptop with access to the internet
- An account on a cloud service provider (we recommend **Google Cloud Platform** (**GCP**); you can create a free tier account here: `https://cloud.google.com/free/`)
- The Terraform CLI tool installed on your laptop
- The Google Cloud CLI tool installed on your laptop (if you are using GCP)

You can find all files for this chapter on the public GitHub at `https://github.com/PacktPublishing/Becoming-a-Rockstar-SRE/blob/main/Chapter08`.

Designing for reliability

Although SREs are not architects, they must have a decent knowledge of solution design and sound patterns. After all, they should be able to identify improvements to the application and infrastructure architecture under the reliability umbrella. Furthermore, they can partner with architects when a new solution is being architected. That way, a solution can incorporate observability and manageability best practices from its cradle.

IT architecture is another profession entirely. It's hard to discuss a vast knowledge domain in a single chapter. We selected the most relevant topics for SREs and divided them into the following four sections:

- Architectural aspects
- Reliability equations
- Design patterns
- Modern applications

Let's start by reviewing the essential IT architecture characteristics.

Architectural aspects

Every time we design a new IT solution, certain aspects must be accounted for and considered. They are called architectural aspects or characteristics and are an integrated part of any drawing for systems.

Availability

Availability is an inherent capacity of an IT system, infrastructure, or application to perform the expected functionalities when necessary. We say a system is available when it can serve a user the advertised services. Usually applied to IT infrastructure only, we extend this concept to the application tier.

A highly-available architecture contains components that work in conjunction to provide the following availability tactics:

- **Fault detection**: Responsible for detecting any faults in the system
- **Fault recovery**: Recovers from a fault condition by replacing the affected component or directing users to another one
- **Fault prevention**: Keeps redundancy constant on the serving and backing components

> **Important note**
> Availability and reliability are distinct domain concepts. A reliable architecture combines reasonable levels of other characteristics such as availability, scalability, resiliency, security, durability, usability, and observability with high-performance systems.

Scalability

Scalability is the ability of an IT system, infrastructure, or application to handle a workload increase without changing its design. Architecture is scalable if it can add IT resources to support additional requests or loads.

With the open adoption of cloud computing, scalability gained scaling approaches. There are four types of resource scaling. Let's see each of them in the following list:

- **Scaling up**: Also known as vertical scaling. It increases the IT resource size to raise the system capacity for load.
- **Scaling out**: Also known as horizontal scaling. It adds new IT resource instances to expand the system's capacity.
- **Scaling down**: This is the reverse of scaling up. It shrinks the IT resources' size to diminish the system capacity for the workload.
- **Scaling in**: This is the reverse of scaling out. It removes the IT resource instances to reduce the system capacity for load.

> **Important note**
> Scalability and cloud elasticity are different properties. While scalability refers to adding (or removing) IT resources, elasticity is the ability to adapt to workload changes by automatically creating or deleting cloud resources.

Resiliency

Resiliency is the property of an IT system, infrastructure, or application to function after a considerable failure at the component level. We say architecture is resilient if it survives unplanned disasters or major failures.

While availability concerns the component level, resiliency targets whole systems and environments. Tactics such as disaster recovery plans, drills, and objectives are common in this arena, and hot-swap or cold-swap backup systems are examples of tactics for resilient infrastructure designs. We will explore design patterns for applications that increase application-level resiliency later in this chapter.

Observability

As an SRE tenet, observability is the ability to determine an IT system, infrastructure, or application state by inspecting its signals and outputs. Architecture is said to be observable if it provides hints and clues about its inner conditions through **metrics, events, logs, and traces (MELT)** data.

We apply the observability principle in the architecting phase to ensure that the solution design captures and incorporates those non-functional requirements.

Next, we understand how reliability is calculated based on the arrangement of components.

Reliability equations

As many of you will know, we intuitively know which component arrangements are more reliable than others. For instance, we know that two components working in parallel as a single system are much better than one in a standalone composition. In this section, we want to explain mathematically why this is true.

Going deep into the mathematics behind the reliability calculation formulae for specific arrangements is always challenging for those who didn't graduate as an engineer. For that reason, we present the equations here without demonstrating their proofs. For more information on the adopted models, check out the *Reliability: Probabilistic Models and Statistical Methods* book from **Leemis**.

Reliability definition

We define the reliability of a single component by its probability of succeeding when serving a request. A reasonable estimate for this probability is to calculate the component's reliable time for a period. Reliable time is when the subsystem attends service requests with a minimum quality. It's the same as saying that reliable time is the serving time minus downtime (when the part was unavailable or performing below the expected level).

Series arrangement

We have a series disposition when we arrange components one after another linearly. For instance, when we connect a web server to an app server that connects to a database in turn, as shown here:

Figure 8.1 – An example of a series arrangement diagram

Imagine a typical e-commerce solution where the user accesses the portal through the web server. It logs into the system that uses a business logic (a Java bean) in the app server to retrieve the user's profile. The Java bean reads data from the relational database and renders the proper layout based on the user's likes. From the user's perspective, the probability of this transaction succeeding is proportional to each component's success. If any of the members fail, then the whole transaction fails.

The reliability index for this system is the multiplication of each probability of success. We give the equation for series arrangement systems here:

$$R_{system} = \prod_{i=1}^{n} R_i$$

Figure 8.2 – Reliability equation for series arrangement systems

Suppose we assume the reliability of each component is 95% for the web server, 96% for the app server, and 97% for the database server based on downtimes for them. In that case, the system's reliability is 0.95 x 0.96 x 0.97, which results in 88.47%.

Parallel arrangement

A parallel arrangement system has components working side-by-side. That means a request can go through any of the parts to succeed. If one of the components is up and running, it will serve the request:

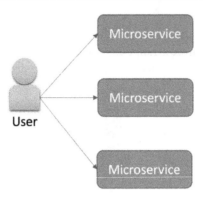

Figure 8.3 – An example of a parallel arrangement system

In the preceding diagram, a user can execute a business transaction through any of the three available microservices. It needs just one to be accessible and serving.

The reliability index for the preceding system is 100% minus the probability of any part failing. We give the equation for parallel arrangement systems here:

$$R_{system} = 1 - \prod_{i=1}^{n}(1 - R_i)$$

Figure 8.4 – Reliability equation for a parallel arrangement system

Assuming each microservice has a reliable time of 90%, the system reliability index is 1 – (1-0.90) x (1-0.90) x (1-0.90), which results in 99.9%. So, you see why we intuitively search for parallelism in IT systems, infrastructure, and applications.

The k-out-of-n arrangement

In a particular case of parallel arrangement systems, we have the **k-out-of-n** configuration considering that k components out of n in total need to at least succeed. This structure is closer to production systems, where a minimal number of IT resources are required to sustain the current workload. In other words, in this arrangement, n-k parts may fail without making the whole system fail. We give the system reliability equation here:

$$R_{system} = \sum_{r=k}^{n}\binom{n}{r}R^r(1 - R)^{n-r}$$

Figure 8.5 – Reliability equation for a k-out-of-n arrangement

This equation assumes that all parts in the composition have the same reliability or probability of success.

> **Important note**
>
> There are much more complex arrangements, such as mixing all the previous types. For such complex structures, we use a tool to apply advanced calculation techniques, such as the **reliability block diagram (RBD)**. This discussion is beyond this book's scope, and you can find more information at this site: `https://en.wikipedia.org/wiki/Reliability_block_diagram`.

In the next section, we will dive into design patterns, which are best practices in architecting resilient solutions for systems, infrastructure, and applications.

Design patterns

Design patterns are the most used general repeatable software or application design solutions that affect infrastructure architecture. They resolve typical problems found during the architecting of a new app. Therefore, SREs must know those patterns, including their limitations, applicability, and benefits for the system's reliability.

Traditional patterns

We will start by looking at the two traditional design patterns that have been used for a long time. The **client-server** and **model-view-controller (MVC)** patterns were the most adopted layouts before cloud computing became the ruler.

Client-server

In this typical design that dominated in the early days of the personal computing era, the software has two distinct components: a client and a server piece. The client component could be a desktop-side app or the native browser connected to a server that had to scale with the number of clients. The reliability of such a design depends on both parts' availability, performance, and the connection between them.

For instance, a user utilizes a client app to access the main application component over the internet:

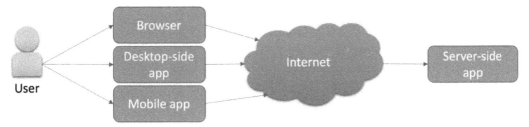

Figure 8.6 – An example of client-server architecture

In a client-server architecture, the client app can be a simple browser, also known as a thin client, running on the user's laptop that connects to a web application hosted in a data center. That imposes a challenge as the web application must be compatible with numerous browsers and client operating systems.

Model-view-controller (MVC)

The MVC pattern splits an application into three parts: model, view, and controller. The *model* piece encapsulates the data model and logic as objects inside the app. In contrast, the *view* is responsible for the user interface and interactions with the app user. And the *controller* is the brain of the app that controls the data flow from model to view, and vice versa. This design pattern was typically deployed to a three-tier architecture with a presentation, business logic, and data layer. This kind of architectural layout is strictly coupled. Therefore, the reliability depends on the scalability and resiliency of the three layers.

A typical MVC architecture is displayed here:

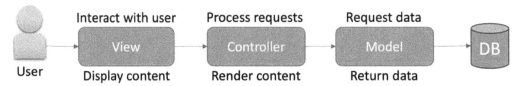

Figure 8.7 – An example of MVC architecture

Each part of this architecture pattern has well-defined and delimited responsibilities, allowing developers to distribute functions and features amongst those components in a straightforward manner.

Microservices

Microservices architecture is the first choice for cloud-enabled or cloud-native applications. It refers to an architectural style where each business functionality is a single-function, self-contained, autonomous, small-scale service, hence microservices. The reason for essentially employing this pattern is that it offers scalability, resiliency, agility, and isolation from failures, which translates to high levels of reliability.

For instance, let's say a user accesses a car rental platform through its mobile phone app. The app will connect to microservices depending on the desired operation:

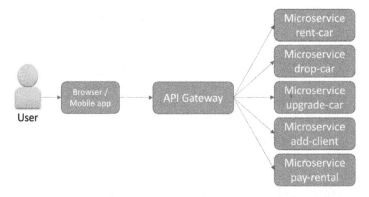

Figure 8.8 – An example of microservices architecture

We use an API gateway (or manager) to properly route the request from the user to the correct microservice since there may be multiple existing microservices, one for each business operation.

Pub-sub

This is also known as the **publisher-subscriber messaging pattern**. A sender (publisher) sends messages that contain events to multiple receivers (subscribers). For example, a topic is a subscription channel where a publisher posts events, and the subscribers read those events. This design pattern is an asynchronous messaging architecture broadly adopted for integrating hybrid cloud environments where systems hosted on-premises must communicate with services running on a public or private cloud.

An example of pub-sub architecture is displayed in the following diagram:

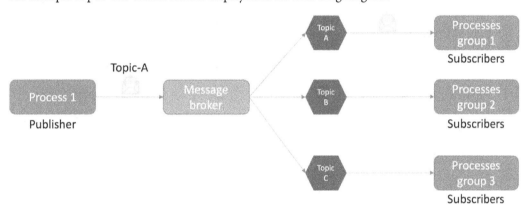

Figure 8.9 – An example of pub-sub architecture

We use a message broker component to distribute a message addressed to a particular topic among all subscribers. Message queues provide buffering for the channels to increase the data transport efficiency. Applications using an event-driven architectural style are examples of pub-sub design pattern adopters.

Suppose there's a requirement for large amounts of data flowing from publishers to subscribers. In that case, apps use the event streaming design where continuous data packets are transferred among the entities instead of discrete messages. This pattern is a much more effective layout when working with media content that needs to be distributed to many clients.

Cloud-native patterns

Cloud-native design patterns help to architect and build reliable apps in a cloud environment. Application designs exploit cloud resources and services for better scalability and resiliency. Each cloud platform provider has a well-architected framework with a recommended set of design patterns, and we will explore the ones that improve the reliability of applications.

Bulkhead

The bulkhead name came from its physical twin used to isolate a ship's hull sections. If the water floods a hull section, it won't go to all areas and make the boat sink. The principle of this pattern is to partition service consumers into pools to isolate failures and avoid an entire cascading collapse. Imagine a client app overwhelming the app microservices with an abnormal request rate. If this app uses a bulkhead design, this client app has a specific assigned microservice instance and will not consume other microservices.

For instance, distinct workloads are served by different connection pools and application instances:

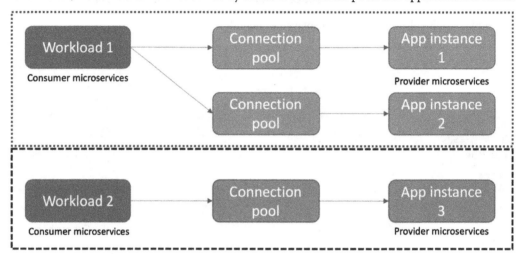

Figure 8.10 – A depiction of the bulkhead design pattern

This pattern ensures that an issue with an application instance will only affect a particular workload and not all the consumers—the bulkhead pattern benefits applications with failure isolation narrowing down the blast radius and boosting overall application reliability.

Circuit breaker

This pattern invokes a remote resource or service through a circuit breaker proxy. The proxy monitors the failures from the remote shared resource or service. If the number of fails is high, it stops passing requests to the remote part and sends back an error to the request originator. After some time, it tries to call the remote part again to see whether it recovered from the fault. Then it will pass all requests to the remote shared resource or service again. This pattern prevents an application from invoking a remote service or using a resource if it's at fault or overloaded and waiting until the error happens.

We illustrate the circuit breaker pattern in the following diagram:

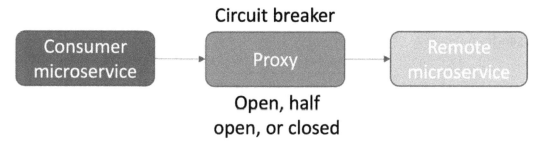

Figure 8.11 – A depiction of the circuit breaker design pattern

The proxy acts as a circuit breaker with three possible states: open, half open, and closed. If the remote service responds appropriately, the proxy keeps itself in a closed (pass-through) state, like a home electrical breaker. If there's performance degradation or failures in the remote service, the proxy moves to an open state (no calls to the remote service). After a set timeout, the proxy is half open and tries to use the remote service again. If it fails again, then it will open immediately. It returns to the closed state when it can call the remote service.

This pattern increases the app's reliability by avoiding long wait times to fail and implementing the fail-fast philosophy.

Deployment stamps

This pattern mandates that we host and operate a set of workloads or tenants onto a group of diverse resources. Each copy of this group, which contains the same resources and services, is called a stamp, and sometimes it's called a service unit, scale unit, or cell.

Although they share the same deployed app, we provision, manage, and monitor each stamp as separate entities. That way, SREs upgrade each stamp's app code or infrastructure resources. Such scale units or cells may be in the same geographical region or multiple areas, but each will handle a pre-determined set of workloads and tenants.

We portray a deployment stamp pattern for multi-tenancy in the following diagram:

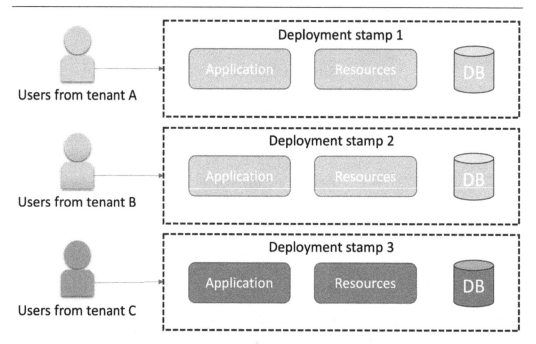

Figure 8.12 – A depiction of the deployment stamp design pattern

For each group of users belonging to the same tenant, there's a group of resources, data, and application instances associated with it. The isolation happens throughout the whole deployment.

Deployment stamps help apps to scale out and isolate any problem to fewer tenants. It also enables apps to work with multi-tenancy requirements where data from tenants cannot be mixed.

Geodes

The **geographical nodes (Geodes)** pattern encompasses the deployment of an application to multiple regions. Each region has a Geode, a group of resources with replicated and paired backing services. Consumers from one area will prefer the closest Geode. However, if necessary, each app instance can serve consumers from any region, as the data should be the same.

In the following diagram, we have an example of the Geode pattern:

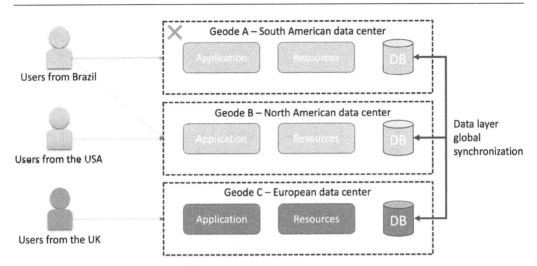

Figure 8.13 – A depiction of the Geode design pattern

Users from a particular country will be primarily routed to the closest Geode based on the geographical distance or network latency. The network directs the users from Brazil to the **South American data center** Geode as it will serve them with the best network delay time. However, if this Geode goes offline, those users are automatically rerouted to the next closest data center, the **North American data center**. Since the data layer is synchronized among all geodes, the Brazilian users' data will also be available in **Geode B**.

This pattern helps apps scale globally and implement near-edge computing to decrease latency, increasing reliability.

Retry

In this pattern, applications handle momentary failures when invoking remote services or accessing shared resources differently. Instead of just capturing the failure details and exiting, it enters a retry-wait loop. This design is helpful when network glitches and problems are common, or the remote service gets overloaded from time to time. One key aspect of the pattern is differentiating a transient from a permanent fault.

The retry pattern is another side of the coin for circuit breakers, and it prevents returning too many errors to the user when temporary errors are frequent.

Throttling

The throttling pattern controls resource or service consumption. When an app adopts this design, it can limit the number of requests for a tenant, user, or another service, to prevent a collapse. It's also applied when autoscaling is taking place to increase resources. While there are not enough resources for everybody, the application throttles the incoming requests according to the preset business rules to minimize the impact.

Many other design patterns are available to improve the application architecture's reliability, and all of them affect the infrastructure. SREs should understand those patterns to work with solution architects for better scalability, resiliency, performance, security, and availability.

Modern applications

Besides the application architecture, how developers structure and manage their code base is also critical. Undoubtedly, microservices with the pub-sub architectural style have been gaining attention under the cloud-native domain. SREs must understand the coding practices around the software development of microservices-based apps.

The Twelve-Factor App

The Twelve-Factor App methodology is a *de facto* standard for microservices apps. According to its authors, it offers the following: "*use declarative formats for setup automation, to minimize time and cost for new developers joining the project; have a clean contract with the underlying operating system, offering maximum portability between execution environments; are suitable for deployment on modern cloud platforms, obviating the need for servers and systems administration; minimize divergence between development and production, enabling continuous deployment for maximum agility; and can scale up without significant changes to tooling, architecture, or development practices.*"

Those statements are precisely what a modern app should offer. We provide a compliance questionnaire in *Appendix B, The 12-Factor App Questionnaire*, to assess a microservices app in terms of factors from the Twelve-Factor App framework.

After learning the basics of creating reliable designs with the design patterns, we check their applicability to efficiently handling workloads.

Splitting and balancing the workload

As SREs, our first order of the day is to ensure we split and balance the workload among the shared resources and backing services. Partitioning the load impacts the reliability as it isolates failures to small sections of the total population. Balancing the load also improves reliability by assigning less loaded resources to attend to a new request, thus decreasing latency. We will understand each concept next.

Splitting

We split the workload based on the consumer profiles as a general practice. We can employ many criteria items to divide incoming requests, but some are frequent in the designs. If the solution is a multi-tenant product, splitting the load by the tenant is a must-have requirement. An independent resources group will serve each tenant. A second example is a global deployment where users are present in many countries. Having the load partitioned by geographical regions is a good practice. Multiple application instances must be available in different parts of the globe. That way, if a particular area faces downtime, it will only affect the users served by the local application instance.

Another way of slicing the workload is looking at the user task level. Depending on the task or operation, it is processed by a specific resource group. If something fails, it only compromises a subset of the application functionalities. For instance, the microservices for **create, read, update, and delete (CRUD)** endpoints related to sales operations run in a different cell than corporate HR operations.

Although splitting and balancing are disparate notions, load balancers do both. Next, we will check out the balancing part.

Balancing

We balance the workload based on an algorithm that understands the conditions of the underlying resources or services. The principle of load balancing is to route an incoming request or user to the next most suitable worker. Adopting this principle brings scalability, redundancy, flexibility, and efficiency to the palm of SRE's hands. And it's considered in all IT spheres, including servers, containers, and serverless environments.

Load balancing algorithms

Let's understand how a load balancing software (or hardware) decides where to route a new request or user:

- **Round robin**: Perhaps the most well-known algorithm, it distributes the load equally to all workers.

- **Least connections**: Redirects a new request or user to the worker with the least number of connections.

- **Least time**: Redirects a new request or user to the worker with the least latency.

- **Hash**: Routes a new request of the user based on a hash code function. This function calculates the next worker based on the originator's IP address or other fields.

Load balancers

The **load balancer (LB)** is a dedicated software function or hardware appliance responsible for balancing the load. There are many vendors and cloud providers that offer such capability. Each of them has its terminology, but in essence, we have four types of load balancers:

- **Hardware-based LBs:** They are the first types of LBs that emerged in the market. Since they have dedicated hardware processors, they can provide an immense throughput.

- **Software-defined LBs:** Commonly present on **hyperscalers**, they are virtual functions inside a cloud platform.

- **Application-level LBs:** Also known as **Level 7 (L7)** LBs, they work with HTTP and HTTPS protocols. Some vendors also offer a **Level 5 (L5)** LB based on TCP or UDP protocols.

- **Network-level LBs:** Also known as **Level 3 (L3)** LBs, they operate on the IP protocol.

After the workload is split and balanced, SREs ensure servers, containers, and serverless groups are reliable enough to attend to all routed requests from the LB. We will learn about the high availability and failover of such resources next.

Failing over – almost as good

Architecture is highly available if it survives a component failure automatically. Either another running component takes over the workload from the faulty part, or a new member is brought to the active state to handle the load. The former is called the active-active scheme, while the latter is named active-standby.

When one of the instances inside a cluster experiences downtime, the LB will stop sending new requests to it. In an active-active mode, the LB will select the next working instance. However, for the active-standby mode, a defect instance triggers a failover process that moves the standby instance to the active state while it tries to recover the in-fault instance.

The failover operation is an intense computing process because it needs to spin off a new instance with the same resources and datasets. Since it takes some time to complete, users will experience an impact during this transition period. For that reason, it's preferable to use an active-active approach. On the other hand, keeping multiple active instances is more expensive as it requires more resources, and data needs to be shared or replicated among instances. Hence, it's not always possible to use an active-active scheme, for example, when we have large datasets that would have a prohibitive cost to keep replicated in two places.

Another approach is the always-on standby architecture. In this case, the standby resource is kept alive and synchronized. Therefore, the time to switch from the active node to the standby one is negligible. A second benefit of the always-on design is that you don't need to automate the failover process, and it's as simple as switching from one node to another.

If the traditional active standby is in place, SREs must first document the failover process as a runbook. Then, they must test the failover process reasonably and automate the entire procedure. The trickiest part of automating a failover is quickly detecting when the active member is under a fault condition. We recommend using a multiple-criteria approach and not relying only on being pinged, as an example. Also, SREs consider transferring authority from one member to another when the script determines a failure. We can apply design patterns to smooth a failover transition to the users.

In the next section, we will discuss what to do when everything functions right, but the workload grows.

Scaling up and out – horizontal versus vertical

Scalability, like availability, is a fundamental architectural aspect that serves as a ground for building reliability. The number of requests or users an application has varies with external factors that are not under control. It's inevitable to find the application demand surpassing the available resources. In such cases, we need to add resources or instances so the application can attend to more requests or users. Adding resources or instances is called **scaling**.

There are two types of scaling, as shown in the following diagram:

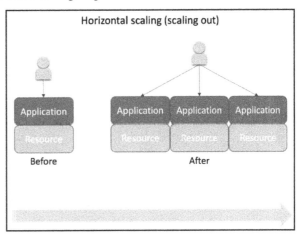

Figure 8.14 – The two types of scaling

There are disadvantages and advantages to horizontal scaling or scaling out, and vertical scaling or scaling up. Also, in a virtualized environment, mainly on cloud platforms, we have the autoscaling feature, which automates scaling.

We will detail horizontal scaling next.

Horizontal

Horizontal scaling or scaling out means adding instances or nodes to an application infrastructure. The LB will have new members to share the workload in a load-balancing group. For example, imagine we have a group of three **virtual machines** (**VMs**) under an LB, and we scale out those VMs if we add a new VM. In other words, we increase the number of workers but not their capacity.

The following list details the pros of horizontal scaling:

- Continuous availability, no need to modify any running VM or node
- No limits due to hardware capacity
- Extra redundancy and resiliency, as each instance is an isolated group of resources

The con of horizontal scaling is detailed here:

- Cost is a concern as new VMs or nodes cost more than just additional CPU allocation

Let's look at the second type of scaling next.

Vertical

Vertical scaling (scaling up) occurs when we add more resources, such as CPU, memory, or storage, to the existing instances. In a load-balancing group, the LB will continue to count on the same number of available members, but those members will have more capacity. For example, imagine having a group of three VMs under an LB. We scale up those VMs if they augment their CPU or memory. In other words, we increase the worker's capacity but not the number of workers.

The pro of vertical scaling is as follows:

- Cost optimization as you're not increasing the number of workers

The cons of vertical scaling are as follows:

- Capacity change is not seamless. You must shut down the current VMs or nodes and bring in new ones with the modified resource requirements.
- Hardware or hypervisor limits as no infinite CPU or memory is available.

Ideally, SREs need to scale up and out if possible. If cost is not a constraint and the app can have multiple nodes, then horizontal scaling has advantages over vertical. In the next topic, we will briefly discuss autoscaling in virtualized environments.

Autoscaling

In virtualized environments, especially cloud platforms, it's possible to use autoscaling features. With autoscaling, the number of workers automatically increases if the conditions of the policy are satisfied, and this completely offloads the burden of automating scalability from SREs.

The Google **autoscaler** implements an autoscaling policy for a group of VM instances. It can autoscale the number of VMs in the group based on utilization metrics:

- Target average CPU utilization
- Serving LB capacity
- Any target cloud monitoring metrics

It can also autoscale based on a schedule. For example, we know we have peak loads at the weekends, so the autoscaler can add VMs during that period:

```
resource "google_compute_autoscaler" "foobar" {
  ...

  autoscaling_policy {
    max_replicas    = 5
    min_replicas    = 1
```

```
    cooldown_period = 60
    cpu_utilization {
      target = 0.5
    }
  }
}
```

The autoscaling policy says that if the target group of instances has more than 50% of all vCPUs used, it will scale out by adding up to 5 VMs.

Let's dive into a practical lab with load balancing and autoscaling in the next section.

In practice – applying what you've learned

To practice what you have read in this chapter, we will outline an autoscaling simulation lab in this section. We play with **Terraform** again, capitalizing on what we have done in the previous chapter's lab.

You will need prerequisite knowledge to completely appreciate this lab, such as the following:

- Familiarity with cloud computing and cloud platforms
- A basic understanding of Terraform

We will divide this practice into three sections:

- Lab architecture
- Lab contents
- Lab instructions

First, let's check out how we have designed this lab.

Lab architecture

This lab uses the HashiCorp Terraform **infrastructure as code** (**IaC**) tool, which relies on the Google Cloud API behind the scenes.

Terraform

It would be best if you have the Terraform **command-line interface** (**CLI**) tool installed on your laptop and the service account key generated for the project inside the GCP account:

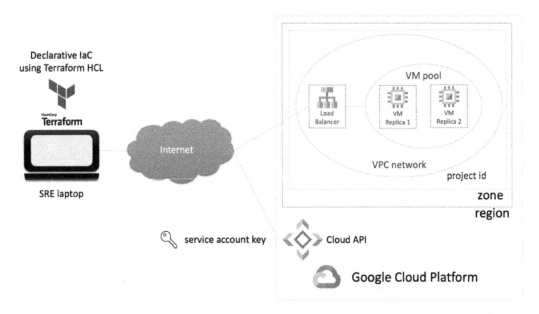

Figure 8.15 – The autoscaling simulation lab diagram

In this lab part, we will create a **virtual private cloud** (**VPC**) network and, inside it, an LB that distributes requests arriving on port 80 to a pool of VMs. This pool of VM replicas has autoscaling capacity enabled in it.

Lab contents

You can clone the GitHub repository by entering the following command in your terminal:

```
$ git clone git@github.com:PacktPublishing/Becoming-a-Rockstar-
SRE.git
```

Within this repository, under the Chapter08 folder, there is just one subfolder called terraform. Also, there's a quick setup procedure called autoscaling-simulation-lab.md in the same.

Terraform

To access the Terraform content, go to this directory:

```
$ cd Becoming-a-Rockstar-SRE/Chapter08/terraform
```

Inside this directory, you're going to find the main.tf file. It contains similar blocks as described in the lab in *Chapter 6, Operational Framework – Managing Infrastructure and Systems*, with the addition of more resources:

The provider block

This block configures the providers by telling them how to connect to the cloud platform:

```
provider "google" {
  credentials = file("project-service-account-key.json")
  project = "autoscaling-simulation-lab"
  region  = "southamerica-east1"
  zone    = "southamerica-east1-a"
}
```

We need to pass the Google service account key, project ID, region, and zone for the google provider.

The resource blocks

On this other block, the resources are described in a declarative manner. They have descriptions of the end state of the infrastructure components:

```
resource "google_compute_autoscaler" "foobar" {
  name  = "rockstar-autoscaler"
  . . .
}
resource "google_compute_network" "vpc_network" {
  name = "rockstar-network"
}
resource "google_compute_instance_template" "foobar" {
  name         = "rockstar-instance-template"
  . . .
}
resource "google_compute_target_pool" "foobar" {
  name = "rockstar-target-pool"
}
resource "google_compute_instance_group_manager" "foobar" {
  name = "rockstar-igm"
  . . .
module "load_balancer" {
  . . .
}
```

Within this block, we declare six distinct resources, as follows:

- The autoscaler policy (`google_compute_autoscaler`)
- The VPC network (`google_compute_network`)
- The compute instance template (`google_compute_instance_template`)
- The target pool (`google_compute_target_pool`)
- The instance group manager (`google_compute_instance_group_manager`)
- The LB (`load_balancer`)

Lab instructions

After you get the `main.tf` file configured with the information from your GCP project, it's time to run it.

Terraform

We assume that you have installed and configured `terraform` here and that you also have a service account key enabled for this lab:

1. Go to the lab directory:

    ```
    $ cd terraform
    ```

2. Initialize Terraform and install any dependencies:

    ```
    $ terraform init
    ```

3. Format the `main.tf` content:

    ```
    $ terraform fmt
    ```

4. Validate the Terraform configuration syntax:

    ```
    $ terraform validate
    ```

5. Provision the declared resources:

    ```
    $ terraform apply
    ```

6. Check the instantiated LB, VMs, and VPC network in the GCP console (optional).
7. Delete all declared resources:

    ```
    $ terraform destroy
    ```

Done!

Similar capabilities are found in other significant hyperscalers such as AWS and Azure. SREs need to adopt autoscaling and load balancing for their deployments for a reliable design.

Summary

In this chapter, we reviewed designing for reliability concepts and how to calculate the system reliability based on each component's success probability. Then, we discussed the modern application considering a reliable design and the Twelve-Factor App standard for them. Later, we explored how to split and balance workloads and why high availability and failover perform with LB. We also reviewed how we scale resources up and out to respond to increasing demand, and which is better and when. Finally, we consolidated the knowledge gained in this chapter by going through the autoscaling simulation lab available in the GitHub repository.

By now, you should be able to explain and use a reliable design and understand how SREs make use of design patterns to check the application architecture's reliability. In addition, you should be able to assess an application structure and code base management using the Twelve-Factor App framework and split and balance workloads. You should also know how to implement high availability by using a failover or always-on approach, and understand how to scale up and out and how they differ one from another. Finally, you should know how to use autoscaling and load balancing as practical knowledge.

In the next chapter, we will investigate ways to detect and eliminate *toil* by applying automation techniques.

Further reading

- To learn more about *Cloud Design Patterns*, please visit the Azure Architecture Center here: `https://learn.microsoft.com/en-us/azure/architecture/`

- To learn more about *AWS Auto Scaling*, please check this website: `https://aws.amazon.com/pt/autoscaling/`

- To learn more about the GCP autoscaler, we recommend the following website: `https://cloud.google.com/compute/docs/autoscaler`

- And for *Azure Autoscale*, please use this link: `https://azure.microsoft.com/en-us/products/virtual-machines/autoscale/`

- You can read more about load balancing in the *What Is Load Balancing?* article from NGINX at this link: `https://www.nginx.com/resources/glossary/load-balancing/`

9
Valued Automation – Toil Discovery and Elimination

In recent years, enterprises of all sizes have funded their business process automation projects. Although the reasons for such investments are decreasing operational costs (CapEx and OpEx) and gaining agility, often the approach adopted for that effort has not been fruitful. A similar situation is found in IT operational work automation, where manual and repetitive tasks are still necessary to keep systems running. Such chores increase the likelihood of error and consume the precious time of engineering teams. **Site reliability engineers** (**SREs**) have a special relationship with automation because of how it affects reliability.

An essential step in site reliability engineering adoption is automating operations and experimenting with coding. However, automation should never be a site reliability engineer's goal, but eliminating *toil* is. **Toil** is a type of work that is repetitive enough to use unnecessary energy from the team but is also automatable.

We'll start this chapter by explaining one of the most vital SRE missions: eliminating toil. We'll look at why SREs should approach an automation requirement as a software problem and how to nail it. Then we'll examine the most automated process nowadays: the **DevOps** delivery pipeline. Finally, we'll go through a few exercises to help practically consolidate this chapter's content.

In this chapter, we're going to cover the following main topics:

- Eliminating toil
- Treating automation as a software problem
- Automating the (in)famous CI/CD pipeline
- In practice – applying what you've learned

Technical requirements

At the end of this chapter, you'll find an automation simulation lab based on the **Kubernetes** platform. We want to ensure you leave this chapter with practical knowledge.

You will need the following for the lab:

- A laptop with access to the internet
- The `syft` CLI tool installed on your laptop
- The `grype` CLI tool installed on your laptop
- The **Docker Desktop** tool installed on your computer (you can create a free tier account here: `https://hub.docker.com/`)

You can find all files for this chapter on the public GitHub at `https://github.com/PacktPublishing/Becoming-a-Rockstar-SRE/blob/main/Chapter09`.

Let's begin with every SRE's purpose.

Eliminating toil

Site reliability engineering disciplines fill the systems management gaps left by the increased complexity of solutions in a hybrid multiple-cloud infrastructure environment. Complexity intrinsically hinders the scalability and reliability of systems by inserting unnecessary burdens in all operations. SREs were born to keep things simple by eliminating repetitive tasks, which is one of their fundamental purposes. To understand how SREs accomplish this mission, we'll divide this section into three parts:

- Toil redefined
- Why toil is bad
- Handling toil the right way

Next, we'll redefine what toil is in the site reliability engineering context.

Toil redefined

Google defines toil as *"the kind of work tied to running a production service that tends to be manual, repetitive, automatable, tactical, devoid of enduring value, and that scales linearly as a service grows."* For a long time, we used this definition to target tasks for elimination. Although it's a good portrayal, we want to expand on it to accommodate hybrid cloud estates and DevOps pipelines.

Toil is the work connected to designing, developing, building, testing, deploying, running, and managing production systems that are manual, repetitive, automatable, and impede growth.

We realized the previous characterization of toil unintentionally limited the definition scope to operational work only, leaving repetitive, manual, automatable, yet strategic engineering tasks behind. Also, SREs don't only play a part in the running phase of production systems; they work in all life cycle steps, including development, building, and testing. With this expanded redefinition, we hope the SRE role gets much more traction in companies adopting it.

Let's understand why manual and repetitive tasks are terrible for the business and SREs next.

Why toil is bad

We say a chore is manual if it needs intervention from a human being, an SRE in this context, either to run a set of instructions or trigger a script. These chores are repetitive if they require multiple executions during a period. Those two aspects make any task time- and energy-consuming, but why is this work problematic?

Manual consequences

First, if SREs must execute manual duties to keep a system running, they will most likely fail at some point. Therefore, manual system management is the antithesis of reliability. The following detrimental results are also consequences of manual tasks:

- **Scalability hindrance**: If the building or maintaining of an application demands T number of manual tasks, then for A instances of the same application, the number of required operations is T multiplied by A ($T x A$). That means the toil will increase proportionally to the size of the environment, which, in technical terms, is an N order of magnitude scale or $O(n)$. No business will survive if we need to add more SREs for each new system deployed.

- **Error-prone**: The more dependency we have on human interventions, the higher the likelihood of inserting errors. And mistakes generate more manual work. Typically, one error unfolds into multiple tasks in the queue on an exponential scale.

- **Technical debt**: It's common sense that keeping documentation is more complicated than code. Procedures or runbooks rapidly become full of technical debt as the knowledge tends to move from the documentation back to the minds of individuals. When an SRE leaves the team, tacit knowledge often goes with that person.

Repetitive madness

Second, if SREs must execute exact instructions many times, are they pursuing innovative ways to make operations scalable? Could they be investing this time in other important tasks? Although repetitiveness shows a stable and standardized process, it leads deeply technically knowledgeable personnel such as SREs to madness. SREs want to explore new paths and not walk in a circle on the same road over and over.

On the bright side, a repetitive task has a great chance of being automated or made automatic. The difference between automated and automatic is subtle. We obtain an automated job by applying an external automation tool to execute the target task on our behalf. We have an automatic job if the software performs the target task as an embedded administrative feature.

Now that we comprehend what toil is and why it's terrible for all, let's check out how to deal with it.

Handling toil the right way

In a nutshell, site reliability engineering teams need to implement three basic processes:

- Identifying toil
- Measuring toil
- Eliminating toil

Identifying toil

Pinpointing manual, repetitive, and automatable chores is like identifying bugs in software code. We have a clear description of what a bug is, and we see the behavior changes caused by it. Still, it's not a trivial job to find all code bugs.

The identification process starts by documenting the primary procedures. Without that, it's not possible to determine what's automatable. SREs spend a reasonable amount of time writing runbooks and standard operating procedures for the most used tasks.

After that, SREs inspect whether a runbook or any straightforward task is toil. This simple check should be part of their sprint retrospective, assuming they work under an agile methodology. If they detect any toil candidate, they must pass it on to the subsequent processes.

Measuring toil

A single squad can identify hundreds of possible instances of toil, but it doesn't necessarily mean the site reliability engineering team will work to eliminate them all. We need to measure each candidate to determine whether it's a priority in the SRE **backlog** and whether the effort is worth the trouble.

The calculation is straightforward. We multiply the number of hours it takes to execute the task by how often it happens per month. Then we divide the previous multiplication by 163 hours per month, giving you a reasonable estimate of **full-time employees** (**FTEs**).

For instance, say a runbook consumes 2 hours for each run, and an SRE must execute this runbook every week or 4 times per month. Applying the formula, this is equivalent to 0.05 FTEs or 5% of an employee's time. We are freeing 5% of human resources if we eliminate it. Not bad!

Eliminating toil

Next in the stack is eradicating the most impactful repetitive and manual chores for the previous reasons. We assume this task is necessary and can't simply avoid it for a particular motivation.

> **Important note**
> As an SRE, you should ask yourself why you are performing any task. Often, no one can provide a decent answer, and the correct action is to then stop executing that task as there's no value.

We have three distinct approaches for handling toil properly:

- Develop an automation script
- Develop an embedded feature
- Optimize the related processes

SREs can develop a code chunk (also known as a script) based on the runbook to automate a task. They can use a variety of programming languages, including **shell script**, **Python**, **Node.js**, and **Golang (Go)**. The resulting code is also called an **automaton**, and any process can trigger its execution. For instance, a job scheduler can run the script at certain times, or a monitoring metric threshold can trigger it.

Another way is to develop a new embedded feature inside the platform or application that automatically takes care of the toil. This strategy is much more efficient than the first one. However, it requires changes to the app's core code, which involves the development and business teams. Many applications and platforms have their maintenance routines already developed as administrative features.

The final approach is to review and optimize the business processes creating manual, repetitive tasks for the SREs. There are many methodologies for that purpose, such as Lean Six Sigma, disciplined Agile, and DevOps.

We will talk more about developing automation scripts or automata next.

Treating automation as a software problem

When we adopt software engineering for automation, we see toil as a requirement and code as its resolution. SREs share this mindset with software engineers: problems are requisites for developing code. Nonetheless, automating tasks through code somehow becomes an even more challenging endeavor. Automation constraints are depicted in the following diagram:

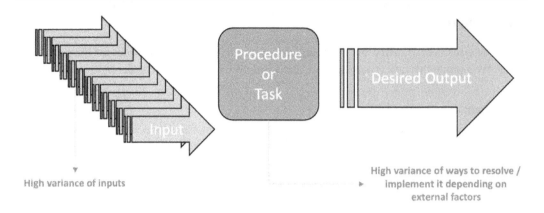

High variance of inputs

High variance of ways to resolve / implement it depending on external factors

Figure 9.1 – Typical procedure or task representation

There are necessary inputs for any given procedure or task to execute it successfully. Also, methods, techniques, or algorithms are applied to such entrance information to provide the desired outcome. This procedure is automatable with simple technologies if only one possible input and output exists. Alternatively, imagine multiple inputs and possible techniques (see *Figure 9.1*). This scenario with numerous possibilities can escalate quickly and request a more robust technology such as **artificial intelligence (AI)**.

Ideally, we address toil at the intersection of automatable tasks, automation tool capabilities, and team software development skills. Here, we will use a pragmatic model, **document-algorithm-code (DAC)**, to streamline the automation enablement process. Let's start with the document phase.

Document

SREs document tasks from either operational or engineering work as detailed step-by-step procedures. They usually share a template that helps put together what they did to resolve or diagnose an issue. It can be the actions necessary to install and configure an infrastructure component. Such standard operating procedures are also called runbooks. They have vital characteristics:

- They follow a model or template. Teams understand how to read and follow them.

- They are visible to the entire organization. Everyone knows where to find them.

- They are valid for the operations (and development) team. SREs and others review them from time to time.

- They contain enough details. There's a balance between maintainability and usability intended for newbies.

We'll describe a good example next.

Standard operating procedure

Task: Deploy a new application

Product: **Internet Information Services (IIS)**

Vendor: Microsoft®

Version(s): 7+

Author(s): Rajesh

Parameters: `name`, `application_pool`, `app_path`, and `physical_path`

Instructions:

1. Open the command-line window as an administrator.

2. Manually add the `inetsrv` directory to the `PATH` environmental variable as follows (so you can execute `appcmd` directly from any location):

    ```
    set PATH=%PATH%;%systemroot%\system32\inetsrv\
    ```

3. Check whether the application already exists. If you see the app in the list that results from the following command, stop the procedure:

    ```
    appcmd list app <name>/<app_path>
    ```

4. Create a new application with the passed parameters:

    ```
    appcmd add app /site.name:<name> /path:/<app_path> /
    physicalPath:<physical_path>
    ```

5. Restart `apppool`:

    ```
    appcmd recycle apppool <application_pool>
    ```

6. Stop the website:

    ```
    appcmd stop site <name>
    ```

7. Start the website:

    ```
    appcmd start site <name>
    ```

This runbook author has included parameters (inputs) and commands (methods) in the preceding example. Next, we'll check how to transform runbooks into algorithms.

Algorithm

Usually, SREs have sharp software development skills and don't need to think about the algorithm and logic before coding a procedure in a programming language. We recommend this intermediary for two simple reasons. First, the SRE team may still be maturing its software engineering skillset, and learning coding by devising algorithms from runbooks is a good practice. We write algorithms using pseudo-code.

Second, if the automation platform changes, it's much easier to port pseudo-code than actual code. Also, it's possible to generate code out of pseudo-code with proper tools.

From the previous section's example, the equivalent pseudo-code would be like this:

```
BECOME Windows admin user
EXECUTE command
    set PATH=%PATH%;%systemroot%\system32\inetsrv\
EXECUTE command
    appcmd list app <website_name>/<app_name>
SET output with previous result
IF output is not empty THEN
  RETURN "Application already exists"
ELSE
    EXECUTE commands block
        mkdir <physical_path>
        appcmd add app /site.name:<website_name> /path:/<app_
name> /physicalPath:<physical_path>
        appcmd recycle apppool <application_pool_name>
        appcmd stop site <website_name>
        appcmd start site <website_name>
    IF any error happened THEN
        RETURN error
    ELSE
        RETURN success
    ENDIF
ENDIF
```

There are other standards for pseudo-code. Check which one is available in the software development tools. The last phase of the DAC model is actually to code.

Code

There's nothing special in this section except SREs must follow the **software development life cycle (SDLC)** processes and tools. They need to adopt the automation platform programming language as well. For instance, if the organization uses **Ansible Tower**, **YAML** and **Python** are the choices for coding.

Next, we'll discuss one of the critical automation targets: the continuous integration and delivery pipeline.

Automating the (in)famous CI/CD pipeline

A **continuous integration/continuous delivery (CI/CD)** pipeline is a process that carries code from development up to production by automatically provisioning infrastructure for building, testing, and deploying an application. It's the core set of technologies that support the DevOps model. Many CI/CD models and automation tools are available on the market, and covering them all would be unmanageable. As SREs, we must understand how a CI/CD pipeline works and what good practices there are to automate it since it's a source of toil. In the following figure, we have a simplified depiction of a **CI/CD pipeline**, also called a **delivery pipeline**:

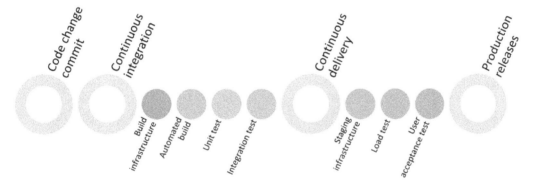

Figure 9.2 – CI/CD pipeline

We'll examine the preceding diagram and consider the following three main sections:

- Continuous integration
- Continuous delivery
- Production releases

We start our journey through the pipeline by understanding how it works to keep all code changes integrated into a single build.

Continuous integration

The trigger for the whole process is a source code change committed to the principal repository. After that, the pipeline needs to provision a build environment where the application build process can run.

Build infrastructure

Provisioning the software build environment is no different than any regular infrastructure provisioning. We can utilize any automation tool to create the required virtual machine, container, or serverless platform where the build tool compiles and packages the code. For instance, if Kubernetes is in place, we can create a script to instantiate a Pod with the necessary libraries to build the source code. We can also apply **infrastructure as code (IaC)** if we work with a cloud-native application.

Automated build

The next step is to run the build automation inside the provisioned infrastructure. This process will pull the source code, compile it into executables, and package the outcomes with other artifacts. If the application runs as a container and uses the **immutable infrastructure** philosophy, we must render a container image together with a **software bill of materials (SBOM)**.

Test infrastructure

Like the build infrastructure, we need to provide infrastructure to run the built application and test it properly.

Unit testing

Automated testing is part of any **CI/CD pipeline**, and this topic alone has dozens of practices. We run a unit test against the application and employ test software or a library to try the application's new feature, functionality, or expected behavior. Depending on the test case, the test algorithm can run in the same infrastructure or a distinct one.

Integration testing

Next in the pipeline is running an integration test for the built software. This test ensures code changes don't affect other modules or parts of the application.

After a successful build and testing, we can delete the provisioned infrastructure components. This de-provisioning practice is also a requirement for **CI/CD pipeline** automation. With the previous actions, we complete the continuous integration pipeline. Let's check the continuous delivery pipeline next.

Continuous delivery

The previous continuous integration pipeline ending triggers the continuous delivery pipeline. We start this new pipeline with a large-scale platform for more robust testing. The objective of the phase is to test whether the application is ready for a production release.

Staging infrastructure

The staging environment should have a similar configuration as the production environment. It contains an old copy of the production datasets as good practice to make the testing more trustworthy. Another automation requirement is to create minimum monitoring for it.

Load testing

After the automation deploys the built software into the staging infrastructure, it starts the load testing. This automatic testing stresses the application and infrastructure limits to determine saturation points. It uses specialized tools to mimic thousands of concurrent requests at the same time.

> **A note for SREs**
>
> We fully implement observability for the staging environment to fine-tune **metrics, events, logs, and traces (MELT)**.

User acceptance testing

The final test is to use real users or bots to test all the application features, ensuring that the application behaves as designed for users. SREs can help in the development of the testing scripts and the deployment of **real-user** or **synthetic** monitoring.

At the end of the continuous delivery pipeline, we have thoroughly built, packaged, and tested software ready for deployment in production. We apply a policy to determine whether to push it to end users or not. The automation decommissions the staging environment if no other adjustments are necessary.

Production releases

If the outcome of the **CI/CD pipeline** is a worthy candidate for production release, the process adds a tag to it. This tag starts the release or **continuous deployment** pipeline. Similar to what we have seen, it deploys the new software version to the existing production infrastructure.

One of the differences with production release automation is that it utilizes a strategy to optimize the deployment while minimizing the impact in case of a defect. For instance, it's common to adopt **blue-green deployments**, **canary releases**, or **A/B testing** for that purpose.

From the automation perspective, a fully automated approach is independent of the chosen production deployment strategy. Not just that, SREs will need to handle automatic rollbacks as well. We explore this topic further in *Chapter 10, Exposing Pipelines – GitOps and Testing Essentials*.

We'll close this chapter with a practical exercise to eliminate toil in the building process in the next section.

In practice – applying what you've learned

To practice what you have read in this chapter, we'll outline an automation simulation lab in this section. We'll use **Docker** to build and package a simple Node.js application. Then we'll combine `syft` and `grype` by adding an **SBOM** to the container image.

You will need the following prerequisite knowledge to appreciate this lab:

- Familiarity with JavaScript and Node.js

- A basic understanding of containers

We'll divide this practice into three sections:

- Lab architecture

- Lab contents

- Lab instructions

Let's check how we have designed this lab first.

Lab architecture

This lab uses `syft`, a CLI tool, and a **Golang** library for generating an SBOM from container images. It also relies on `grype`, a vulnerability scanner for container images. It assumes you have a **Docker Hub** account set up.

It would be best to have the `docker`, `syft`, and `grype` commands installed on your laptop. You can find more details about installation at the following links, respectively:

- `https://docs.docker.com/desktop/`

- `https://github.com/anchore/syft#installation`

- `https://github.com/anchore/grype#installation`

In the following diagram, we work with the Docker Hub as the container image registry. Use `docker login` to authenticate with your username and password to create a Docker repository. After that, clone the Packt GitHub repository:

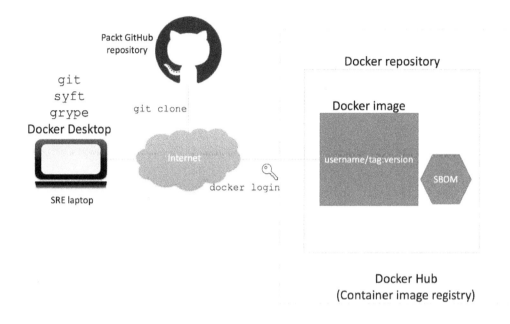

Figure 9.3 – Automation simulation lab diagram

We will create a container image from a Node.js application using the docker command, then use syft to generate this image SBOM, and finally, scan it with grype. We will automate this build process with a simple shell script.

Lab contents

You can clone the GitHub repository by entering the following command in your terminal:

```
$ git clone git@github.com:PacktPublishing/Becoming-a-Rockstar-
SRE.git
```

Within this repository, under the Chapter09 folder, there is just one subfolder called sbom. Also, there's a quick setup procedure called automation-simulation-lab.md in the same directory level.

SBOM

To access the SBOM content, go to this directory:

```
$ cd Becoming-a-Rockstar-SRE/Chapter09/sbom
```

Inside this directory, you're going to find the `Dockerfile` file. It contains instructions for the Docker build tool to assemble a container image with the Node.js app located in the same directory:

```
FROM node:16
WORKDIR /usr/src/app
COPY package*.json ./
RUN npm install
COPY server.js .
EXPOSE 8080
CMD [ "node", "server.js" ]
```

More details on the Docker build instructions and arguments are available at this link: `https://docs.docker.com/engine/reference/builder/`.

We placed an automation script in the same directory to run all steps to build a container image, attach an SBOM, and check for vulnerabilities. It contains environmental variables in the first block:

```
#!/bin/bash
export VERSION="0.1.3"
export USERNAME="rod4n4m1"
export APPNAME="node-api"
export PASSWORD="xxxxxxxx"
```

The second part contains the Docker commands to build and publish the container image in the Docker Hub registry:

```
# Build the application container image and push it to the
repository
docker login -u $USERNAME -p $PASSWORD
docker build . -t ${USERNAME}/${APPNAME}:${VERSION}
docker push ${USERNAME}/${APPNAME}:${VERSION}
```

The last part creates an SBOM and scans it for vulnerabilities:

```
# Create an SBOM and attach it to the image on Docker Hub
syft ${USERNAME}/${APPNAME}:${VERSION} -o spdx-json > sbom.
spdx.json
# Scans the SBOM for known vulnerabilities
grype sbom.spdx.json > vulnerabilities.grype
# TODO
# cosign the SBOM and attach it to the repository
```

As an additional challenge, check the `cosign` tool for signing the SBOM to guarantee it came from the originator.

Lab instructions

It would help if you changed the following environmental variables in the `app-build.sh` file accordingly. They need to reflect the values for your **Docker** username and password:

```
export VERSION="0.1.3"
export USERNAME="rod4n4m1"
export APPNAME="node-api"
export PASSWORD="xxxxxxxx"
```

After you get the `app-build.sh` automation script configured with the information from your Docker project, it's time to run it.

SBOM

We assume that you have installed and configured the Docker CLI and Docker Desktop here. Also, you must have a Docker account enabled for this lab:

1. Go to the lab directory:

    ```
    $ cd sbom
    ```

2. Run the automation script:

    ```
    $ ./app-build.sh
    ```

3. Done!

We found similar capabilities in other automation platforms, such as **Ansible** and **Jenkins**. For the sake of learning, we used the most straightforward approach possible. SREs should eliminate toil by having sharp skills in automation techniques and tools.

Summary

In this chapter, we understood what toil is and what it is not by redefining it slightly differently from the original Google definition. Repetitive and manual tasks are always burdensome from an SRE's point of view, and SREs must identify, measure, and eliminate them whenever possible. Automation is an elegant answer for automatable toil, and resolving it as a software problem is a good practice. We dissected CI/CD pipeline automation. Finally, we consolidated the knowledge of this chapter by going through the application build simulation lab available in the GitHub repository.

After reading this chapter, you should be able to explain toil and automation through an SRE's eyes and how SREs detect and eliminate manual, repetitive, and automatable chores that can drain energy and inhibit growth: documenting runbooks and scripting them for automation. You should also be able to explain how the CI/CD pipeline works and the automation requirements for it, and know how to use an automated application build process, as practical knowledge.

In the next chapter, we dig into the DevOps pipelines and continuous deployment, embracing a new philosophy called **GitOps**.

Further reading

- To learn more about **runbooks**, please visit the PagerDuty® article here:

  `https://www.pagerduty.com/resources/learn/what-is-a-runbook/`

- To learn more about **CI/CD pipelines**, please check this website:

  `https://www.redhat.com/en/topics/devops/what-cicd-pipeline`

- To learn more about **cloud-native CI/CD**, we recommend the following website:

  `https://tekton.dev/`

- And for the **automation platform**, please use this link:

  `https://www.jenkins.io/`

- You can read further about load balancing in the *What is Pseudocode? How to Use Pseudocode to Solve Coding Problems* article from freeCodeCamp at this link:

  `https://www.freecodecamp.org/news/what-is-pseudocode-in-programming/`

10

Exposing Pipelines – GitOps and Testing Essentials

The thought of GitOps being part of an SRE book may be shocking to some given that GitOps is an essential part of the DevOps movement – and you would be right. But this isn't just any SRE book; rockstars see the great opportunity to embed potential reliability and observability into a pipeline.

GitOps started when a developer decided to use automation to run a script against their code when taking an action such as merging code in. In fact, GitOps is loosely defined as automation that happens based on triggers in a Git repository. While the most popular use case is the automatic compiling and deploying of code and infrastructure, we can leverage GitOps to test code, interact with internal processes, and even trigger deployment failure due to the lack of unit test coverage in a repo.

GitOps has also caused a rise in popularity for **Infrastructure as Code (IaC)**, bringing code and infrastructure into our pipelines. This not only allows us to troubleshoot problems with code and infrastructure but also gives us the capability to inherit an infrastructure's properties to build monitoring and alerting. Alongside the popularity of IaC, the industry is moving toward a newfound capacity we call **Monitoring as Code (MaC)**, a topic that will be well explored in this chapter.

The focus of this chapter is both providing a simple overview of pipeline functionality and understanding how pipelines can provide an extremely proactive SRE response to prevent outages and reduce developer toil.

We are going to cover the following main topics:

- A basic pipeline – building automation to deploy infrastructure as code architecture and code
- The automation of compliance and security checks
- The automation of code quality and standards using linting
- The automation of testing to validate functionality during deployment

- The reduction of developer toil through the automation of process

- In practice – applying what you've learned

A basic pipeline – building automation to deploy infrastructure as code architecture and code

GitOps systems that deploy pipelines are scripting engines that allow complex routines to run. These can be as simple as a single JSON, YAML, or bash script, or extremely complex, looping in other repositories, templates, and even external API calls. The basic ideals of pipelines are remarkably similar between vendors, even though the syntax may be different.

A typical pipeline will leverage variables in its decision-making for us. The pipeline system will introduce a number of pipeline variables, such as what directory the code is in, the hash of the latest commit, the repository name, and even protected values pulled from internal or external protected data storage engines.

The next essential part is the stages, or steps, of the pipeline. These can be used to compile code, deploy IaC, or even perform testing and scan code for malicious content. These blocks often are unable to talk to each other directly and should be considered independent steps in the pipeline processing.

Finally, we define actions, often as Linux scripting inside the pipeline stages. The build process is often separated into before, during, and after steps, making it possible for us to break up deployments – for example, installing dependencies or removing cached files would be found in a before script.

Pipelines in chronological order

Pipelines steps exist in a timeline (see *Figure 10.1*). As you can see, the stages can have multiple items of work within them, which are called in chronological order. If one of these items should fail, the processing typically stops, commonly referred to as **breaking a build**.

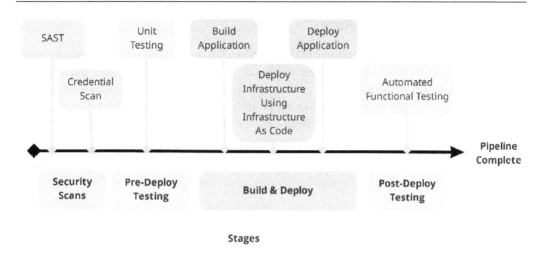

Figure 10.1 – Basic pipeline diagram

When a build breaks, many of the actions taken cannot be undone. For example, if the build breaks in the **Post-Deploy Testing** stage, then the application has already been deployed – and because there is no *rollback*, or undo, for deployment, this deployment will continue running in the environment it was deployed to.

Pipeline templates

One of the most powerful features a rockstar SRE will leverage in pipelines is the use of templates. Whether found in a system-defined *templates* repository or named from another repository, the idea is simple – make reusable pipeline tooling and scripts to share between multiple repositories and pipelines.

This power allows us to rapidly deploy changes and features to an entire group of repositories in one go. This could be in the introduction of an API call to push data from repositories that capture standard information about deployments into a central storage system, or even entire deployment routines.

The standardization of these pipelines gives us the ability to not only introduce new features quickly by adding a few lines to the repositories or altering existing templates but also, by using the same code again and again, test the templates in repositories with little impact should they break, gaining trust to use them in mission-critical deployments.

Errors or breaks in pipelines

Anyone who runs pipelines will inevitably see one fail. The reasons for pipelines failing range from simple code compilation issues and files being referenced in the wrong directories to complex IaC issues and the addition of testing and validation steps.

Using containers in pipelines

We often leverage containers when building pipelines. Why? The simple answer is these give us clean build environments with no files left over from past builds. We can also introduce tools and libraries into the container we run the build in, which can reduce compile time by not having to download these items when the pipeline runs. In fact, we often use repositories to build the containers, which are used to build other repositories! And yes, we can build a deployable container inside another container.

Containers also provide another great avenue for adding automation and even deploying tooling such as APMs, logging aggregators, and security tooling.

Pipeline artifacts

Often, there is a desire to save specific files for review in case of issues. These files can range from simple security and testing reports to logs or even entire directories of files from a process. SREs often use verbose logging in pipelines to see the highest level of detail in logs and store the log file in artifacts for rapid issue resolution.

Caution! Care should be taken when keeping artifacts from production. Artifacts are often not stored securely, and if production credentials are saved as artifacts, those credentials may easily be accessed. It is always best to think of artifact storage as unencrypted and insecure storage.

Pipeline troubleshooting tips

When the build breaks inside of a pipeline, most of the time, it's an issue with compiling the code, or an issue with your IaC. There will be occasions, however, when finding what broke your build eludes you at first glance. When I look at a failed pipeline, I'm looking for very specific things in both the logs and the files inside the repository. We'll review a list of the common areas I look at now.

Troubleshooting with logs

Not only does the GitOps pipeline often generate logs but you can also set up applications such as **Terraform** to increase the level of logging and even debugging. These detailed logs can provide insight into why pipelines failed.

Logging can also often be enabled by setting an environment variable to a specific value, for example, setting the TF_LOG variable to the value of TRACE will tell the Terraform binary to enable logging at the most verbose or detailed level. Since it's an environment variable, this can simply be added as an option to the build or as a quick additional line of code in the pipeline – without the need to make any further adjustments.

Troubleshooting with artifacts

If we have artifacts saved, it gives us visibility into the files created during the build process, especially when your build processes fetch files that are not present in the same format in the deployment binaries. A great example of this is software libraries, which get compiled into applications.

Troubleshooting with directory listings

By far, the most common issue with pipelines I've seen in my career is the expectation of a file being in a specific directory or location, only to find out it's not where I thought it would be, or I forgot to add a specific prefix to the file, perhaps `. . /` or `. /src/`, which gives an exact reference to where a file is. This can be difficult, as we may switch directories to build something or save files fetched from other systems into directories other than where we would expect them to be.

Simply using a directory listing, `ls` in Linux/Unix or `dir` in Windows, and allowing the files to show up in the build logs will greatly impact your ability to see which files exist and where these files exist. This trick is by far the best remediation tactic for issues with files not being found in pipeline builds.

Troubleshooting out-of-order events

Just because a pipeline breaks in one area does not mean the problem is in that area. Let's look at an example of this (see *Figure 10.2*). The pipeline defines a variable in the **Build Application** phase and then uses it in the **Deploy Application** phase.

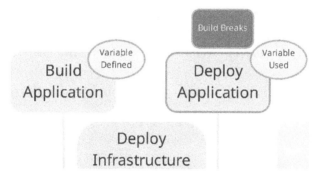

Figure 10.2 – Pipeline variables and an out-of-order break

In the diagram, you can see the build breaks when the variable is used. Upon closer investigation, we find the problem is that the variable itself is not defined properly.

As there is dependency in most pipelines, an issue in one area may not be visible until that dependency is used later in the pipeline.

Troubleshooting by displaying variable values

So, what do we do when we want to review the values inside a variable set in an environment? Simple, we use a command to display that variable. This simple command would display the value of the SAMPLE variable:

```
echo $SAMPLE
```

This statement will provide an output of the variable's value and add it to your pipeline log. As with artifacts, the value held in the SAMPLE variable is not encrypted or secure, so be mindful of what you are logging. For this reason, we should also refrain from using commands that show all of the environment variables.

Troubleshooting by commenting code

The final troubleshooting mechanism I often use is to comment out code in the pipeline in an attempt to find the source of an error. Similar to the *breakdown and testing* troubleshooting method, we simply comment out blocks of code until we have a successful pipeline run. Once it runs, add back in commented-out blocks until you find the source of the issue.

As an SRE, at some point, you will find yourself troubleshooting deployment issues with pipelines. These troubleshooting tips represent the majority of repetitive issues many of us see in our careers and can provide quick paths to remediation.

Two of our greatest abilities are providing feedback and compliance tooling using automated pipelines. Next, compliance and security automation can provide support for enforcing standards such as tagging the architecture created by IaC and the standardization of security scanning.

Automating compliance and security in pipelines

Compliance and security in line with multiple standards are required in many lines of business. The progression of compliance and security scanning has been a natural progression of the rise in popularity of pipelines, given their versatility, extensibility, and proximity to code. A rockstar SRE isn't particularly interested in the scans themselves; however, the generated report can provide insight not only into the age of the code but also into the quality of its upkeep.

Library age

When SREs look at a code base, the age of the libraries being used can be an indicator of the upkeep of the code. As time progresses, often, operating systems, including those that are serverless, and containers are updated. I have witnessed libraries in production applications that were 5 or even 10 years old – and had lost compatibility with newer versions of operating systems and container images. In fact, if the container image version increases, old libraries can cause outages due to compatibility.

Application security testing

It is common to scan applications in two manners. Dynamic testing is applied from the external world, often against APIs and websites to identify exploits both in the architecture and code. Meanwhile, static testing involves scanning code in its raw state, to identify known vulnerabilities; for example, a known security issue with a specific version of an open source library.

Dynamic Application Security Testing (DAST)

DAST tests applications by accessing the application from the outside, typically to attempt to break the APIs used. This type of testing, while critical to security, can provide great insight into how the system reacts under high load as well as attacks. Rockstar SREs can leverage high load testing such as DAST to keep an eye on code quality issues, such as memory leaks and database slowdowns due to the load.

In addition, DAST scanning can cause a high error response while in process. This can trigger alerts and cause confusion if teams are unaware the scanning is happening. One way of preventing this is by using a static IP source for the DAST scanning and filtering those calls out of HTTP access logs.

Static Application Security Testing (SAST)

SAST looks for known vulnerabilities in the code itself, providing feedback not only for vulnerable code but also for poor code usage techniques, which can lead to security vulnerabilities. The biggest task of SAST today though is the analysis of known libraries for flaws. As libraries of code, especially in the open source world, are known to have vulnerabilities, SAST provides immediate feedback to the user if errors are built into a pipeline. You can even fail builds if they have unmanaged vulnerabilities.

Secrets scanning

Secrets scanning is a more recent type of scanning that scans your application code base for passwords and secrets that are stored in your code. A simple rule of security is that no passwords or keys should be stored in code. Why is this important to an SRE? Simple, organizations that provide their development teams with production credentials often see an increase in the amount of ad hoc production – the most common impact this can have is malformed or slow queries to production databases, which can cause table to lock for long periods of time. Other issues can also occur, such as using production systems for testing alongside development or QA environments, which can produce a wide range of issues.

SREs should understand the source of all traffic going to production systems. If production systems are not being accessed correctly, there may be a higher failure rate for the system.

Security, like reliability, provides a positively focused tension on delivery that should be expected and welcomed. Providing this feedback directly and with the clarity a pipeline brings has an amazing impact as a guide rail for securing applications.

Beyond just security and compliance, the automation of the pipeline grants us other capabilities. Next, we'll discuss what linting is and how it can provide feedback to developers about code quality and implementation.

Automated linting for code quality and standards

Linting is a tool that provides direct feedback on improving both the quality and readability of your code. Based on a rule set, linting provides valued feedback on how code is written. Unlike security testing, which looks for known exploits, linting rules can identify poor coding habits that could be problematic, for example, infinite loop detection. And yes, linting provides feedback on standards in variable naming and code spacing, creating highly standardized and easier-to-read code. Of course, code can compile and work in a not-so-readable state – but like grammar in the English language, code should be clean and easy to understand, not a garbled mess.

Beyond just style, linting can provide feedback on a number of possible scenarios, including the following:

- Code that will never be executed
- Possible situations where indexes will be outside their array's bounds
- Code and data that may not be in a format that can be ported well to other OSs
- Possible infinite loops
- Unused variables
- Poor code formatting
- Common potential syntax errors

Often, we find that flags raised by linting can be wild goose chases, which simply eat through development time and have little to do with actual functionality. For a rockstar SRE, understanding the deficiency of the underlying code provides great feedback on the team's code quality and possible items to investigate when doing code reviews.

Compiling with linting feedback

When Google commissioned the creation of the Go language, one of their basic requirements was to provide developers with immediate feedback and even refuse to compile code with issues such as unused variables. In short, the Go compiler has linting built in and refuses to compile code if the code does not meet a specific level of quality.

Why is this unusual? It's one of the first compilers to take into account syntax and style and force developers to be more responsible with their roles as developers. The Go compiler will also rewrite the source code to resolve formatting issues. Many other compilers will issue warnings for items such as unused variables, but the Go compiler is one of the few, in its default configuration, that refuses to compile software because of them.

Beyond providing valued feedback to developers, linting allows a team to follow *like practices*, which is both invaluable to bringing new members up to speed and can also increase developer output. In a world where discussions exist about the formatting of variable capitalization and tabs versus spaces, the value of having a tool that provides that guidance based on a set of rules is that it allows developers to focus on the functionality of the code instead of style. We've discussed security, compliance, and linting as parts of an automated pipeline; next, we'll dive into test automation.

Validating functionality during deployment with automated testing

The joy a developer gets from seeing something work for the first time can be all-consuming. With the feeling of satisfaction often comes a drive to build more, do more, and yes, be even more clever with our code.

Modern systems are becoming more complex than ever, relying on multiple microservices, databases, and even third-party systems. The ability to test every single possibility by hand has become near impossible. With this complexity, it's not difficult to see why automation has become so highly favored. In an ecosystem where code is developed by individual developers and then merged into a larger code base with multiple inflight development efforts, automated testing ensures that the code, when brought together, still functions as intended.

As testing has progressed from manual point-and-click testing into automation, we also find gaps in many QA engineers understanding of the methodologies of development and infrastructure testing automation. This deficiency greatly impacts both development time and the impact of outages.

Why is testing so important to reliability?

A rockstar SRE is as concerned about preventing outages as they are about resolving them once they happen. In a world where outage management is extremely reactive, testing is one of the tools used to be more proactive in preventing alerts.

Proactively testing the functionality of systems before bringing them to production allows us to provide feedback to the development teams regarding how well the code ran on the system as a whole. In the case of nightly regression testing, we can provide that feedback within a day of code changes that have imparted flaws in a system. In addition, unit testing can provide feedback near instantly to developers. The faster that feedback gets to developers, the closer the developer is to changing that code, which can accelerate development in general.

For SRE, testing is the one true preventative method we have in our arsenal.

Test data

Most testing simply would not be possible without data that allows the system to emulate production usage, for example, a test user account. This test data in itself can be problematic. This issue is further compounded by multiple environments, such as development and QA.

When working with test data sets, we may need access to other environments in order to create data to use for testing, especially when that data is destroyed during testing.

The types of testing

We see a large variety of testing methodologies and each has its value. As an SRE, of course, the concern here is with toil – especially regarding manual point-and-click-based testing, where human error can cause a test failure. We'll explore multiple testing types that can be performed both prior to deployment and even when development is still in flight.

Manual point-and-click

Manually having a person test a system is considered both the most expensive method and the method most prone to errors. Because the tester only has to know the system in which they are performing tests, the test engineer does not require a development background. Also, the manual nature of the testing makes it easier to retrain others with very quick ramp-up times. However, manual point-and-click testing is the least reliable testing method due to the human element, making this the least effective and often most costly testing type over the long-term timeline of a project.

When the testing cost over years is considered, automation provides the best value, despite requiring automation engineers who cost much more than simple manual QA engineers. Because automation can also run more often, we also catch development issues faster and can provide a more comprehensive test set overall.

Automated or synthetic testing

Functional testing provides a faster, repeatable way of testing software. Automated functional tests of APIs provide simple and easy interfaces to make API calls and compare them to expected results. In addition, headless browser technology allows us to actually simulate a user visiting our website and performing actions.

Unit testing

Unit testing is the testing of individual units of an application. Typically written by the developers, a unit can be anything from a small function to even an entire class. We often use mock-ups, or fake data, to perform unit testing. In basic unit testing, we call a unit to validate its output.

A great example of this is a simple function that adds two numbers. The unit test will call the function with 1 and 2 and expect a return value of 3. If the return value is 4, the test will fail. We may also look for test failure as part of testing; for example, 1 and null, depending on how the function is supposed to work, could return 1 or null, or even an exception.

Unit testing is often done with a unit testing framework. Some of the more popular ones are Jest and Mocha. Unit testing mechanisms can exist inside of the development environment to allow you to run a unit test without even leaving the same screen you're writing code on, making unit testing fast. They can even run inside a pipeline before compiling and deploying code. Some frameworks may even calculate how much of the code is covered by unit testing.

One downside of unit testing is the usage of mock or fake data, so this testing should be accompanied by automated functional testing.

Lightweight or liveliness health checks

While not technically part of testing, health checks are used once in production. The health checks are simple checks typically used for containerized applications; when the health check fails, the container is then replaced. It should be noted that health checks do not include checking the state of dependent or downstream applications or databases.

When a call is made to a lightweight health check, only the health of the system itself is checked:

Figure 10.3 – Lightweight health check

When deploying new code, the pipeline can check the health check and decide whether the container is up or down. Based on that test, you can choose to fail a deployment, and even roll back the container version to a prior version.

Heavyweight health checks

Lightweight health checks only validate that the simplest of functionality is available, such as the ability to return an HTTP 200 response code. Heavyweight health checks also include checking downstream systems, such as databases or other APIs.

When a call is made to a heavyweight health check, the system's health is checked and then its dependencies; when all dependencies are checked, a single status is returned:

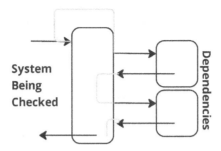

Figure 10.4 – Heavyweight health check

Heavyweight health checks are often tools used for observability and troubleshooting, but rarely for automation.

When to test a pipeline

A pipeline's timing is important, and an SRE wants to test as much as possible before deployment, but we are limited to linting and unit testing. Automated functional testing is done after deployment.

Testing observability

Observability is not just limited to applications and the infrastructure it runs on. With the high value of testing in a highly reliable system, understanding not only the state of QA but also the history of how testing has progressed is vital.

There are several different test tracking platforms, including QTest, though I prefer to have my test results sent as metrics to observability systems such as New Relic or Datadog. Not only does this allow me to build dashboards showing both the current state and trending state but I can also set up alerting for nightly regression testing or even send alerts based both on testing and the performance of applications.

Automated rollbacks

When we start adding testing and validation to pipelines after deployment, when a pipeline deployment fails, we need to tear it down and restore the prior version. This can be as simple as deploying the prior container or as complex as rolling out entirely new clusters of servers, and doing this from within a pipeline adds more complexity.

Blue/green and *canary* are deployment methods in which the old and new versions of an application run on independent copies of infrastructure. For blue/green deployments, we can simply switch traffic from an old version to a new version of an application. Canary deployment allows for a percentage of traffic to run on the new version of an application while being monitored for errors. Should either type of deployment see errors in the new version of an application, restoring the old version happens very

quickly. In both cases, this method can automate returning traffic to the old version of an application, which is called rolling back. Implementing either adds complexity both to the deployment and rollback mechanism, which now takes on external monitoring. The complexity of rolling back is best addressed with architecture that has rollback capabilities built in. We also often limit the time frame during which a system can be rolled back to mere hours after deployment.

This type of deployment technique is amazing to see in action – and it has the added benefit of automating the rolling back of software. Next, we'll discuss reducing developer toil using automation.

The reduction of developer toil through automated processes

In the world of engineering, we often perform the same actions over again and again. We see this type of work every day – a systems engineer deleting temporary files from a server to prevent hard drives from filling up, or a developer manually uploading files every time an application is deployed. These items, which are repetitive, can often be automated, saving time spent on manual tasks and human interaction – when this happens, we reduce developer toil.

When we look at our pipelines, we see example after example of the reduction in toil. By adding SAST security scanning, credential scanning, and unit testing to a pipeline, these functions run automatically. If they fail, we simply break the build – and then a developer finds out why:

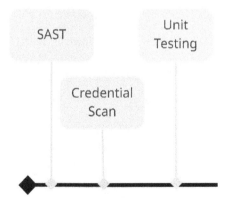

Figure 10.5 – Examples of pipelines addressing toil

If we repeat this process again and again, for everything from writing scripts to retrieving logs from a server you need or even refreshing data in a test database, we reduce toil. We can even automate these processes – and simply be alerted when automated systems have or find issues.

What is the impact of addressing toil?

Toil is seen as a time thief, often involving boring tasks – or as a friend of mine refers to it, *the eating of broccoli*. Most people don't enjoy toil and see it as a duty, rather than something that provides them with long-term satisfaction like writing a new software program does. Because they are manual tasks by definition, they are also subject to human error. When we automate tasks, we reduce the chance of human error and eliminate the drudgery of human toil.

In practice – applying what you've learned

In this lab, we will walk through a simple pipeline that deploys a single microservice in **Amazon Web Services (AWS)** using the AWS **Serverless Application Model (SAM)**. The AWS SAM is a solution that allows for the rapid creation and deployment of both code and architecture for serverless, **AWS Lambda**-based applications. Once you have configured the pipeline to run in your own GitHub account, we'll look at strategies using which we can increase reliability and reduce toil.

GitHub is the most popular Git repository platform today and drives open source and developer collaboration in ways we could not have even imagined before it first came online in 2008.

As a pre-requisite to this lab, you will need the following accounts:

- A free GitHub account: `https://github.com/join`
- An AWS account: `https://aws.amazon.com/free`
- Basic knowledge of using Git and basic file management skills

In addition, you'll need a working Golang development environment consisting of the following:

- Visual Studio Code: `https://code.visualstudio.com/download`
- Golang: `https://go.dev/doc/install`
- Git: `https://git-scm.com/downloads`
- The AWS **Command-Line Interface (CLI)**: `https://aws.amazon.com/cli/`
- The AWS SAM: `https://docs.aws.amazon.com/serverless-application-model/latest/developerguide/install-sam-cli.html`

Preparing AWS for the lab

We start by setting up AWS so GitHub can deploy to your account. This is done by creating an IAM user with the correct permissions to deploy and generate an **access key ID** and **secret access key** – think of these as a username and password. If you are new to AWS, I suggest setting up **AdministratorAccess** and deleting the user in your AWS account at the end of the lab. Note that your secret access key will only be presented to you on the screen once; after it leaves the screen, you will never be able to retrieve it. Make sure you save these credentials:

```
https://docs.aws.amazon.com/IAM/latest/UserGuide/id_users_create.
html#id_users_create_console
```

Creating your repository

For this lab, you will create a new GitHub repository and download a zip file to populate your Git repository with. Once you have pushed the files to the repository, the pipeline will run and attempt to deploy the application to AWS.

To start the lab, you'll need to create a new GitHub repository in your GitHub account using this link: `https://docs.github.com/en/get-started/quickstart/create-a-repo`.

Adding secrets to your repository

In GitHub, we have the ability to retain secrets in each repository. This allows us to store items such as a username and password, or an access token, in secret. You can access these secrets in your actions; however, they will not be printed in the logs. This also allows us to release source code without having our account information in the code.

For this lab, we need to store the AWS access key ID, secret access key, and region in GitHub secrets. To do this, we navigate to the **Settings** tab in our GitHub repository. Then, select **Secrets** on the left-hand side, and choose **Actions**.

To add an action, click on **new repository secret**, and enter your secret name and the secret you want to store. You'll create three secrets:

- `AWS_ACCESS_KEY_ID`: The access key ID from your AWS IAM user.
- `AWS_SECRET_ACCESS_KEY`: The secret access key from your AWS IAM user.
- `AWS_REGION`: The region of AWS you want to run in; you can use `us-east-1` if you are unsure.

```
            HASH: ${{ github.hash }}
            REF: ${{ github.ref }}
        run: |
            # Set variables
            BRANCH=${REF#refs/heads/}
            REPOSITORY=`echo $REPO | tr "/" "-"`
            #This is used for the bucket name so, it must be
compliant with naming conventions
            # https://docs.aws.amazon.com/AmazonS3/latest/
userguide/bucketnamingrules.html
            ENVIRONMENT=$REPOSITORY-$BRANCH
            # In this step we are setting variables and
persistenting them
            # into the environment so that they can be utilized
in other steps
            echo "branch=$BRANCH" >> $GITHUB_ENV
            echo "repository=$REPOSITORY" >> $GITHUB_ENV
            echo "environment=$ENVIRONMENT" >> $GITHUB_ENV
            # Output variables to ensure their values are set
correctly when ran
            echo "The region is ${{ secrets.AWS_REGION }} (shows
as *** because it's a secret)"
            echo "The repository is $REPOSITORY"
            echo "The environment is $ENVIRONMENT"
            echo "The branch is $BRANCH"
            echo "The hash is $HASH"
            echo "The ref is $REF"
```

You will also see we can use an `echo` command piped to the `$GITHUB_ENV` variable to define the variables in the `run` section. In addition, we output the values of the variables; this allows us to see the values of the variables for faster troubleshooting.

Next, we build the application. Notice the `go get .` command – this calls the Golang compiler to fetch the dependency libraries required to compile the application:

```
- name: SAM Build
    run: |
        cd hello-world
        go get .
```

```
            cd ..
            sam build

        - name: Run unit tests
          run: |
            cd hello-world
            go test
            cd ..
```

And of course, we run the tests on the code to validate that the code is functional before we deploy it.

Finally, we define the S3 bucket name and deploy the application using the AWS SAM CLI:

```
- name: SAM Deploy
      run: |
          # Create S3 Bucket to store code
          BUCKET=${{ env.environment }}
          #make bucket name lower case
          BUCKET="${BUCKET,,}"
          echo "using bucket $BUCKET"
          aws s3api head-bucket --bucket "$BUCKET" 2>/dev/null
\
            || aws s3 mb s3://$BUCKET
          # Run SAM Deploy
          sam deploy --template-file .aws-sam/build/template.
yaml \
            --stack-name $BUCKET \
            --s3-bucket $BUCKET \
            --parameter-overrides \
              'ParameterKey=Name,ParameterValue=example \
              ParameterKey=Version,ParameterValue=${{ steps.
vars.outputs.version }}' \
              --capabilities CAPABILITY_IAM CAPABILITY_NAMED_IAM
```

Adding more steps

As you can see, the pipeline is simply a sequential list of steps building one after another. Finally, to expand the pipeline, you simply add additional steps. Let's say you wanted to add DAST or SAST testing; you could simply add a step to call `securego` (from `https://github.com/securego/gosec`) as follows:

```
- name: Checkout Source
     uses: actions/checkout@v2
  - name: Run Gosec Security Scanner
     uses: securego/gosec@master
     with:
        args: ./...
```

This simple step can easily be added by an SRE wishing to increase the security of an application.

Testing but not deploying

When you are working with development branches, you can also simply test your code without deploying it by running steps only on specific branches; for example, if you were to add this line to the `SAM Deploy` step, the step would only run on the `main` branch:

```
if: github.ref == 'refs/heads/master'
```

Lab final thoughts

Like asking someone to build a program, you can take many paths to provide the same functionality when building pipelines. And pipelines can do more than just deploy software; in fact, I wrote one to generate the zip file you downloaded for this lab – which reduces my toil by having GitHub generate the zip file anew every time someone pushes items to the main branch.

Summary

The subject of GitOps can fill entire books. As rockstar SREs, we learned to leverage pipeline automation to meet our goals. We can reduce toil by automating not only deployments but also security and compliance scanning, testing, opening change requests, and providing developers with feedback, enabling them to iterate faster.

The true value of the pipeline is testing. We talk about making SRE proactive instead of reactive all the time, looking for ways in which we can prevent downtime – instead of responding to an issue at 2 a.m. – working to end the expense of half a dozen or more people on an outage call, which steals time away from development. Testing is the one proactive tool that absolutely works.

In addition, we walked through a GitHub Actions pipeline to illustrate just how simple it is to add additional steps and hopefully brought you some real understanding of pipelines.

Next up, we'll discuss serverless deployment, containers, and Kubernetes – the worker units of the cloud as it is today – including what the rockstar SRE needs to know to ensure the highest stability and reliability.

11

Worker Bees – Orchestrations of Serverless, Containers, and Kubernetes

The one common thread in all of IT is that we implement processes that do work. From looking up information about an order that used to be done by thumbing through paper files or signing up for a library card, we leverage computers to not only make life easier and faster but also more secure. And as processes have moved to computers, it's gotten more complex – so complex, in fact, that processes such as approving a simple credit card charge are just not possible without them.

The priority and throughput of completing work will always impact how we coordinate doing work. Does it need to be done now? Do I need more workers to get it done faster? Do I need spare workers in case a worker fails? And when a worker does break something, how do I ensure their work is still done?

This chapter is all about how we leverage today's technology to perform work. From containers to server farms, serverless technologies, and even orchestrations such as Kubernetes, we have many methods available at our fingertips. We'll explore each mechanism for deploying compute technology capable of performing work, complete with the pros and cons shown in real time, and how we marry architecture with code to increase worker reliability.

In this chapter, we are going to cover the following main topics:

- The multiple definitions of serverless
- Containers and why we love them
- Kubernetes and other ways to orchestrate containers
- Deployment techniques and workers
- Automation and rolling back failed deployments
- In practice – applying what you've learned

Technical requirements

This chapter will cover the topic of compute, which is easily defined as a piece of technology that runs code. A basic understanding of programming and web-based technologies such as APIs is required. In addition, the in-practice exercise utilizes Gitpod and GitHub, along with a requirement of a basic understanding of file management and Git.

The multiple definitions of serverless

The term **serverless** in the industry can be nothing short of confusing. In its most basic definition, it's the deployment of resources in a way that abstracts away the server. Serverless provides only the service you need and hands over control of the underlying server to the cloud provider.

There are a number of advantages and disadvantages with serverless. In terms of server management chores, such as operating system patching, log rotation, and even security hardening, serverless hands these over to the cloud provider, reducing staff workload. Serverless is often also considered to be scalable in its native form, meaning it grows and shrinks with load automatically. The biggest disadvantage can be the cost but only at extremely high volumes of calls – and that's a different discussion. The opposite is also true; when you have low volume, serverless can also provide functionality at a fraction of the cost of running servers.

However, the fan-favorite feature of serverless is simply how fast development and deployment is. You simply say, "*I need this,*" and BAM!... you have it.

Serverless Framework

The **Serverless Framework** is a development tool used to deploy applications and infrastructure into the cloud. As a framework, its power is the simplistic mixture of both code and infrastructure into a single unit, deployable in a matter of minutes. Under the hood, the framework builds the required **Infrastructure as Code (IaC)**, interacting with the cloud provider to deploy both architecture and code.

The Serverless Framework is maintained by Serverless Inc, with assistance from the open source community; you can read more about the framework at `http://www.serverless.com`.

Serverless computing

The idea of serverless computing is not new; many services in the cloud are architected from the ground up to provide a service – even though on the backend they utilize a server. From parameter and secret storage systems to NoSQL databases and even logging, alerting, and monitoring, systems are, by their nature, serverless. These services run without the need for us to define the underlying hardware – in other words, the hardware underneath being used is not known or even of concern to the service or user.

In the past few years, serverless computing has grown to include more traditional server-based systems, such as database servers, which are being rolled out in serverless fashion, allowing for even more hands-off operations.

But the undisputed king of serverless is the serverless function, also known as **Function as a Service (FaaS)**. If you hear the word *serverless*, it's often referring to serverless functions.

Serverless functions

At the heart of the Serverless Framework are serverless functions, with the most popular being **Amazon Web Services (AWS) Lambda**, although closing in on it fast are **Azure Functions** and **Google Cloud Functions**. All have similar functionality and costs are often quoted in millions of transactions, making them extremely cost-effective.

Serverless functions can be configured with different memory sizes and timeout limits. The amount of CPU is often a function of memory configuration, so the more memory a function is configured with, the more CPU it's given when it starts. Billing is typically calculated as a function of memory and the length of time the function runs.

Serverless functions start on demand and can be triggered through a number of different methods more numerable than we have time for in this entire book – but one of the most common triggers is via a gateway that turns a lambda function into a web service, either to serve entire web pages or act as an API.

Because serverless functions start on demand, when demand goes up, so can the number of functions – allowing for a high amount of scaling capacity for your applications, inherently automatically within the architecture.

From a rockstar SRE perspective, serverless functions provide many benefits of reliability, including a simple lack of hard drives to fill up, or servers to reboot. Often, this type of architecture narrows down issues to a limited list:

- Cloud provider issues, which can include not only the serverless functions' capability but also issues with the underlying systems that the serverless architecture depends on, such as the cloud storage system that a serverless image is stored on. Often, an outage at the cloud provider level is not acknowledged for at least 20 to 30 minutes.

- Downstream dependencies, ranging from databases and other APIs to storage, can cause serverless functions to fail.

- Bugs in the code.

When we look at this list, the value of QA testing and monitoring of downstream services is apparent.

Beyond just being scalable and inexpensive, serverless functions often support containers, allowing us to merge the benefits of both. Keep serverless functions in mind as we talk about containers next.

Monitoring serverless functions

You would think since we have no hard drive to fill up, or servers to reboot, monitoring serverless would require less effort than our more traditional applications running on a server. While it's certainly different, the level of effort remains the same. The only hope for reducing effort in monitoring is IaC or automation.

Invocations

The simplest items to monitor are **invocations**, or the action of running serverless functions. Understanding if, when, and how often your serverless functions run can provide invaluable insight.

Memory

Since one of the billable factors of a serverless function is the amount of memory it needs, keeping the memory setting as low as possible saves on cost. However, if the memory setting is too low, the function can crash.

When we look at memory usage, we want to ensure that the memory matrix is not the configured memory but the actual memory used for a serverless function. The actual memory used may not be available in the metrics and may have to be extracted from logs. An example of this is AWS Lambda, which is one of the most popular serverless function offerings available today; its metrics give you the amount of memory provisioned, or set up, for the function – but you must look in **AWS Cloudwatch Logs** at each serverless invocation for the amount of memory actually used. You can, of course, set up an **AWS CloudWatch log filter** to create a metric from the logs.

Timeouts

The amount of time a serverless function runs is often configurable as well. When this time frame expires, the function is forcefully terminated. Platforms often implement hard limits as to the amount of time a serverless function can run as well – for example, AWS Lambda is limited to 15 minutes.

We need to be careful as well with a cloud provider's metrics relating to timeouts as well; often, the metric is defined as the provisioned or configuration value, not the actual time the function ran. Again, this limitation can be overcome in AWS by creating an AWS CloudWatch log filter to generate metrics from the logs.

When you see timeouts for large, longer-running functions, some options to mitigate the issue include splitting the functionality into multiple functions and invoking one function within another, using a queue to delegate work to multiple functions and, of course, converting a serverless function into a different architecture, such as containers.

Oddly, because some cloud providers provision CPU as a function of memory, increasing the memory setting on a function can provide more CPU processing power – for CPU-intensive operations such as encryption functions and image or video processing, it's a possible remediation technique.

Concurrency and throttling

Concurrency or the ability to run multiple serverless functions at once is a simple and very effective way to scale out a service under load. And while we would think that serverless functions are capable of scaling to handle millions of requests a minute, those types of numbers are not possible in many basic cloud computing accounts.

Two different types of concurrencies arise in the cloud, the first being **provisioned concurrency**, or the ability to say we need x number of functions ready at all times. The second is based on cloud computing account limits – or a cloud provider simply saying, you must consciously ask to use high amounts of resources.

Throttling is the term used for when a serverless function is prevented from starting more copies of itself, due to reaching a concurrency limitation. When this happens, the functions you expect to be running, say for a web API, will simply not run – in the case of a web API behind a load balancer or API gateway, you may simply get a 500 error.

Skynetting is another popular term you'll hear around concurrency and throttling; it's a movie reference from *Terminator* in which *Skynet* is the artificial intelligence that expanded nearly endlessly and took over the Earth from humans. From an SRE perspective, this term indicates architecture is expanding unreasonably, likely due to a human error in code or an uncontrolled load request. Put simply, skynetting is the expansion of your systems, well beyond what you intended, as if it had a mind of its own.

And finally, **account limitations** set in place by cloud providers are often present to prevent both unexpected costs and the occasional instance of skynetting. Often, these are set to very reasonable amounts and simply require a request to be submitted to be increased.

Errors

Technology sometimes has very simple and clean answers to questions – an error, however, is often of the messiest and most difficult areas an SRE works with. Not only is it common to find teams calling items errors when they are not, but even simple industry-standard HTTP return codes are also often misused. As we've seen with memory and timeouts, getting to the heart of counting errors is not as simple as looking at a single serverless function metric.

Errors, also known as exceptions, in serverless functions come in two basic types, **caught** and **uncaught**. Most development languages have tooling that allows a developer to capture an error or exception and provide additional logic to take action when these occur – we call this a caught exception. Uncaught exceptions are errors that simply terminate the running application with an error message.

These two different errors, caught and uncaught, present differently when we look at serverless function metrics. Because caught exceptions are handled in code and never reported back to a serverless function's subsystem, they are not represented in the *error* metric available in the cloud provider's system. Thus, caught exceptions are errors that occur but are not represented in the error metric.

Since caught exceptions often generate a log of an error, we can look to our logs to generate a count of caught exceptions, using something similar to an AWS CloudWatch metric filter on the logs, and generate metrics from the logs. This gives us a second metric of the caught exceptions. When using this technique, we also want to ensure the caught exceptions have a very standard format, which is identified when filtering exceptions into metrics.

So now, we have two different error or exception metrics, caught and uncaught. In most alerting systems, however, it's easy enough to add these two metrics together to generate alerts. We should also consider that uncaught exceptions provide an area of improvement for developers, in the form of additional code-based exception remediation.

Externally visible errors

When serverless functions are utilized as a part of a service called by other parent services, we can expect to see errors being presented back to the parent service. In the case of APIs, a 500 error is typically returned. We can also receive errors not even related to the code inside the serverless function.

Authentication or permissions and incorrectly configured architecture issues in the cloud can prevent the invocation of serverless functions, resulting in errors being returned to the parent service – without associated logs or error metrics in the serverless function.

While serverless is implemented in many different forms, the idea is the same – deploy a program without needing to thing of the underlying physical hardware, efficiently, and in a highly reliable environment. Next, we bridge the gap from serverless to the world of containers. By bringing together the world of the OS and code into a single, deployable unit, containers have revolutionized technology even more than serverless.

Containers and why we love them

Physical servers were the start of the computerized world we know today and are still widely used. However, running workloads on entire servers can be cumbersome and, worst, make it difficult to share resources, such as memory and CPU, among multiple processes to utilize hardware to its highest potential.

Even before the traditional *servers* as we know them today were created, multi-user, room-filling platforms created by the likes of IBM, HP, and Sun Microsystems allowed us to scale both memory and CPU in a single environment. Processes were run with limitations on resources, which later gave way to virtualization and containers in the late 1990s and early 2000s. As the need for processing power increased, so did the desire to run multiple workers independently of each other, and we started to standardize the way we use hardware – in essence, abstract the hardware away and just request the resources to run a process.

In 2013, when **Docker** emerged as the most complete ecosystem for container management of its time, the popularity of containers skyrocketed and Docker led the containerization revolution.

What has a container become today? In its most basic form, a container is a complete, self-contained OS that you can add applications and files to – in essence, an entire computer in a single **container**.

Isolation

Now, everyone has suffered a computer crashing in front of them, whether a laptop or desktop. You end up reaching for the power button to manually shut down and restart the computer. In a virtualized or containerized environment, processes are isolated from each other, allowing one to even completely crash or use its entire allocated CPU, memory, or disk space without crashing other containers on the same physical servers. Imagine independence and containment running on an entire data center, where each program can be isolated from the others.

This isolation is what makes it possible not only to start and stop containers on a dime but also to replace faulty or crashed containers. Since each container is separate, multiple identical containers can run on the same server or be spread, or **orchestrated**, across many servers without interfering with each other's operation.

Did I mention that containers can be deployed and run in phenomenal amounts of time? Because we bypass the installation of software and operating system updates, we don't have to copy configuration files or compile and load software. In fact, container deployments taking over a minute are considered slow! Nothing beats deployment speed!

Immutability

Containers can be built new each time you want to use one. This allows us to **pull** a container from a source, the most notable being `https://www.dockerhub.com`. After the container is pulled, you can run it, or even use it as a base to further add applications, configuration files, and even data to – and save the entire augmented container again for reuse.

Why is this so valuable? The immutable nature of containers allows us to freeze all aspects of the environment that the software runs in, from the OS to applications and configurations, even support software used for tasks such as security scans, logging, and web proxies. By freezing all these items, we can ensure they remain the same every time we deploy them.

Promotability

Without containers, code is often built and deployed to servers, where **Quality Assurance** (**QA**) teams test the functionality. When tests pass, we again rebuild the software and deploy it to production. But if something is different on the servers, or even if a library is updated between the time you deploy to QA and the time of production, you can have a separate, new version of code running in production that was never fully tested.

With an immutable container, we gain the unique ability to test the container in a QA environment. After we test the container, we can then simply promote it to production, rolling out the exact same OS, applications, configurations, and even data embedded inside the container.

Tagging

Tagging, or the assignment of a simple **key-value pair** to architecture, gives us the means to add metadata to describe an object. This could be as simple as a human-readable name or contact person, or as complex as the git repository URL used to generate the infrastructure and software.

Tagging is also used on containers to identify the differences among the same containers built differently. The most common use is tracking updates to the container or internal software version. Containers can have multiple tags; the most common is the hash of the last git commit used to build the software but can include items as simple as date/time or application IDs from internal application registries.

There are a few tags, however, that are reserved. The most common of these is **latest**, which is used to indicate the most recent version of a container. It should be noted here though that future release candidates and testing versions of a container may exist that are actually newer than the container tagged *latest*.

Rollbacks

When we deploy a container to production, we also typically store old versions of the container – and if something is found to be wrong in the new container, we can roll back to the old version.

This rollback capability gives us the ability to restore old versions of software extremely quickly and in ways that reverting code commits just can't match. In addition, new and exciting deployment techniques such as canary deployments are now possible, thanks to the ability to have containers of multiple versions at our fingertips.

Security

While nothing can be entirely secure, because containers contain both an OS and software, we can scan more than just the software. In fact, multiple container storage systems or repositories perform security scanning as part of the storage process – giving you instant security feedback.

While containers are immutable, unfortunately, security scanning of containers is not. Long-lived containers need to be scanned at regular intervals because new security issues are being discovered every day that could impact previously well-rated containers.

Signable

To enhance the trust and security of containers, we should mention that containers themselves can not only be scanned for security but signed. Signing allows us to validate that the source of a container is a trusted one. This prevents someone from injecting a container with malware into your registry and causing problems.

Monitoring containers

The monitoring of containers can grow quickly into not only monitoring the containers but also monitoring the orchestration of the containers. We start with the basics of memory and CPU, which, unlike serverless functions, are separate and operate on the same tangibility that a simple server would. And we monitor containers much like we monitor servers – although containers have some automation in place that makes their operation more hands-free than servers.

Container restarting

Why would we worry about a container restarting? Containers restart for a number of reasons, including when containers move from server to server to rebalance orchestrations across server farms, or when a server is being decommissioned.

But when containers continuously restart, there is often some reason that needs to be looked into. Be it an unhandled exception terminating a program within the container upon certain conditions, or simply running out of memory, the continuous restarting of containers is something to be watched for in either metrics or logs, and an alert is often built to notify when this is happening.

Now that we understand what containers are and why they are so useful, we can dive into how we leverage containers together in orchestrations to build simple services and large multi-service clusters that run entire companies.

Kubernetes and other ways to orchestrate containers

 Imagine a world where a construction foreman is able to instantly replace a worker when one gets hurt or doesn't show up for work. How much more efficiently could the construction work be done if we could instantly add electricians or repurpose plumbers into roofers. Work would certainly be done in a more predictable and plannable way.

This freedom to replace, increase, and even reassign assets such as memory and CPU is why container orchestration solves so many work cases in the industry.

Health checks

Health checks play one of the most vital parts in orchestrating containers – the ability to ensure a container is running in good health. By asking the container if it's okay every few minutes, we continuously know the state of our workforce.

Containers often implement an HTTP endpoint that returns a simple HTTP 200 response. The endpoint may return simple information such as free memory, software versions, or even the amount of work that has been done. Health checks should never make calls to downstream services such as dependent APIs or databases. Should one of these dependencies fail, the inclusion of them in the health check can cause cascade failures of entire systems and make them extremely difficult to retrieve.

Orchestration services will repeatedly continue to call the health check endpoint. Should the endpoint fail in a timeout or return an unfavorable HTTP response, such as an unauthorized 403 message, we would consider this a failure. It is rare that a system is set up to destroy and redeploy a container should one failure occur; typically, we set up the system to require multiple failures (2 to 5) in a small period of time (often minutes), before we replace a container with a new one.

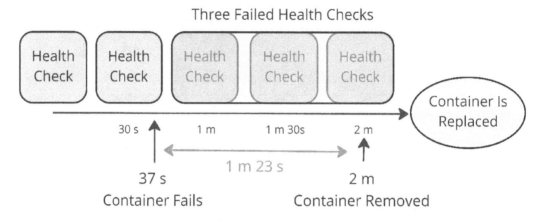

Figure 11.1 – Container health checks and replacement

As we can see, the container fails at 37 seconds, but it takes 1 minute and 23 seconds before three health checks fail, and the container is replaced. During this time, we allow traffic to still flow to the container, although it will fail. This will increase our failure rate during this time. The defense against this time of failure is the orchestration of multiple containers and retries. The retries have a high chance of hitting a different container when they make them retry a request. This retry mechanism is at the heart of highly available systems interaction.

Crashing and force-closing containers

Health checks are not the only ways to trigger replacing a container. Containers can also crash or be forced to exit. Whether due to bugs in code, memory leaks, or even poor error handling, containers can crash on their own. When this happens, the container orchestrator detects the crash, either by an exit code or the failure of a container's health check, and replaces the failed container. Because containers can also be restricted to a specific amount of memory or drive space, when these limitations are breached, the container will also crash and be replaced.

The rockstar SRE is always on the lookout for containers that continue to fail over and over again, referred to as **flapping**. When containers are flapping, we absolutely want to know. We should also point out that the restrictions to memory and drive space can be what we call soft, or not cause a container to crash when going above its limits, and this, as you can imagine, can be bad – very, very bad.

HTTP-based load balancing

While orchestration provides a path to running multiple containers, we must still be able to accept requests from a single source – say, a single URL – and distribute it across the entire fleet of containers. Load balancing can be done in many different ways:

- A round robin where requests are simply sent to all containers in order – for example, 1, 2, 3, and then again 1, 2, 3

- Least connections, which route requests to the container which is currently responding to the least requests

- Least response time, which routes requests to the container that is responding fastest at that moment

An additional setting exists called **sticky**, which allows repeated requests through a load balancer to return to the same container, based on a cookie or IP address. This unique setting allows containers to maintain state or use a cache local to each container.

Server load balancing

Server load balancing uses many of the same concepts as network and HTTP-based load balancing, except it's balancing the load of containers running on a number of servers. Because containers can have inside small microservices or even large monolith programs, balancing them across multiple servers provides not only a way to utilize as much of the server as possible, but also a way for the load to be sent to other containers if one server should fail, which gives you a highly available service.

Now that we've discussed the basics of orchestrating containers and balancing a load across both servers and containers, as well as how container health is handled, we'll dive into a few different orchestration techniques.

Containers as a Service (CaaS)

Just like serverless functions, containers can be run as serverless. This technique is simple; a container or multiple identical containers are simply given a place to run. Load balancing and health checking may not even be available in this type of environment.

CaaS can be very useful, especially in SRE. When we eliminate *toil*, we may want to run a script to update some files, or compile data to export – these small functions can be wrapped into containers and deployed as workers without the need for servers.

Simple container orchestration

When we start wanting to run a variety of containers on a fleet of servers, or even run containers in serverless architecture, simple container orchestration systems can provide powerful reliability without the complexities of Kubernetes. Among this type of service would be **Amazon Elastic Container Service** (**ECS**), which allows multiple containers to run both on AWS Fargate CaaS offerings and virtual servers with ECS deployed on them, with the advantages of health checks and automation to connect to load balancers.

The word that should be remembered here is *simple*. The idea is to allow containers to run well supervised without much complexity. This simple container orchestration is often used for web API services and queue processing-type tasks.

Kubernetes

Built by Google, **Kubernetes**, also known as **K8s**, is a highly configurable orchestration platform and, quite simply, one of many ways to orchestrate containers.

One of the biggest advantages of Kubernetes is its ability to internally network between other containers, both external and internal to physical servers, without the need to leave the server and reenter via the load balancer. This capability alone can greatly reduce response time across an entire multi-service transaction and add a level of networking security often not available in other types of orchestrations.

Kubernetes internal networking can also be governed by a number of different network stacks or implementations – some written specifically for tasks, such as balancing load across servers in different physical locations, providing high levels of redundancy.

It's also worth noting that Kubernetes can orchestrate not only containers but also a number of different virtual environments. And because Kubernetes has built-in persistence of storage options, even databases can be deployed to Kubernetes.

Deployment techniques and workers

The goal of deployments is simple – make changes to environments with the least amount of impact. As an SRE, we are always looking to improve deployment techniques, and the focus in recent years has been on the **Zero Down Time** (**ZDT**) deployments that can happen at any time.

Traditionally, deployment time frames were a compromise between a business not wanting to impact customers and staff availability. As you can imagine, early morning Sunday deployments are a favorite in the industry, due to Sunday being the least impactful time for customers, but it's not a favorite time frame for most staff.

Traditional replacement deployment

In a traditional deployment, current containers are simply forced to quit and new ones are made to replace them. This simple and effective technique may take down an environment for a few minutes, as containers are replaced by your orchestrator.

Rolling deployment

If you take a traditional deployment and replace one container at a time with a new one, you get rolling deployments. This allows you to keep the environment up as long as you have multiple containers running.

One note on this type of deployment – if your environment is under load, it is common practice to first scale up the number of containers. This is because you will be taking down single containers one at a time, and if the remaining number of containers cannot handle the load, you can cause an impact on customers and revenue during deployments.

A/B or blue/green deployment

A/B or blue/green deployments leverage the use of a load balancer or domain name service entry changes to change traffic flow from one set of containers to another. This type of deployment requires you to have two complete sets of containers up and running at once, an A and B or blue and green set, allowing you to switch to the new containers and, if need be, switch back to the old.

While this type of deployment has the highest infrastructure cost, it's also the only one that allows us to instantaneously change load from old to new containers and back if required. And when paired with CaaS, that cost can be limited to only running both for perhaps a day and shutting the old set down, once confidence has been found in the new set.

Canary deployment

Canary deployments allow a small amount of traffic to be sent to new versions of software. This has long been a favorite deployment technique of leading technology companies, providing fast paths to production, but the requirements for doing a canary deployment, if ignored, can cause a disaster.

Deploying in the canary style is rather simple; you leave most of your traffic on the current version of the software and split off a small amount of traffic to send to the new version. Then, you use observability to watch the small amount of traffic to ensure no errors occur. Once the new code has proven itself to be error-free, the percentage of traffic flowing to the new version is increased until all traffic is now going to the new version.

This type of deployment requires a higher level of observability than we normally find and is often coupled with automated rollbacks so that when errors do occur, we can remove the new version from an environment – that is, roll it back.

Having learned about how to deploy containers, we will now see how to leverage container health, service metrics, and rollbacks to quickly remove newly deployed containers from production if they are defective.

Automation and rolling back failed deployments

Now commonplace in most software development shops, the automation of deployments is one of the leading strategies now in place due to the new DevOps movement. When we hark back to deploying software by manually copying files, it's easy to see why this toil has been so eagerly rendered down to automation.

Rollback metrics

Rollback metrics can be extremely difficult to generate, as they have a number of requirements, which can include the following:

- The ability to determine the different versions of source code creating the metric
- Being able to mark the source code version as new or old
- The removal of metrics generated by testing

Separation of the metrics by the code version is best done by tagging the hash of the `git commit`. This will allow each version of code to stand alone in monitoring, but since the ID is just randomly generated, it does not show which version is new or old.

New or old versioning of code is not easily resolved. Tagging *new* and *old* on versions, or the *latest* tag, is ineffective, because at some point, the latest is not the latest, and new becomes old. We can use the total amount of traffic separated by version, and if your canary deployments roll out less than 50% of the traffic to the new version, then switching all traffic to the new version will work.

The most effective way I've found to target new versus old code when defining rollback targets is through the use of monitoring, which is deployed by script or IaC. In this method, we deploy the monitoring infrastructure every time, allowing the source code to inherit the hash of the commit. By performing the deployment of alerting infrastructure every time, we control what is considered new and old, every time we deploy.

Finally, be aware that testing and security scanning, which can generate errors, should be excluded from monitoring related to rollbacks. Ultimately, you may simply need to account for this error volume when calculating just how many errors will cause a rollback. Just remember that as testing grows, so may these error counts.

When to roll back

Rollback decisions can be tricky – especially when working with systems that are in a state of error. The basic idea is to detect issues in the code by subjecting it to everyday traffic and seeing what happens; should errors be generated, we remove the new code.

Defining that error rate though can be difficult. And you may not want to deploy using a canary methodology during low-volume parts of the day, especially for monolithic code that performs a variety of functions. Defining the period of time to deploy is certainly an **Service Level Agreement (SLA)** discussion topic that should be discussed with your business partners.

In determining when to roll back, we want to ensure we describe the **Service Level Indicators (SLIs)** and **Service Level Objectives (SLOs)** and understand both the nature of what we are measuring, and the levels at which a rollback will be triggered. This could be as simple as a 0.5% error rate across an application, or as complex as defining multiple error rates across a number of different functional areas of code. The time frame is another consideration on when to roll back, as we don't typically want rollbacks to happen days after deployment.

How to roll back

How to roll back is highly dependent upon your architecture. This is typically easier in an orchestrated or serverless function environment, as versioning of containers and serverless function code can provide nearly instant reinstatement of prior versions of code.

We want to attempt to leverage the technology available to us in performing rollbacks, and we try to avoid situations where we manually have to rebuild or don't have seamless methods of rolling back, such as manually replacing code on a server.

Rollbacks should also be near **Zero Time Deployment (ZTD)** operations, with little to no impact on incoming traffic or functionality.

In practice – applying what you've learned

This lab emulates what happens when a single container fails in production and the ability of retries to allow dependent services to continue functioning with little to no impact. As a treat, we'll also introduce Gitpod, which is a containerized development workspace that makes development fast and simple and, more importantly, demonstrates just how powerful containers are.

Leveraging Gitpod – a containerized workspace

In celebration of our love of all things containerized, we will be introducing a new tool, **Gitpod**. This workspace tool embeds an open source version of Visual Studio Code along with pre-installed tools, such as a Golang compiler, and provides you rapid access to git repositories, including integrations for GitHub and GitLab.

Start by signing up for a free Gitpod plan at `https://www.gitpod.io/pricing` and installing the Gitpod browser extension from `https://www.gitpod.io/docs/configure/user-settings/browser-extension`. Then, navigate to this book's GitHub repository and open this repository immediately in Gitpod by clicking the **Gitpod** button:

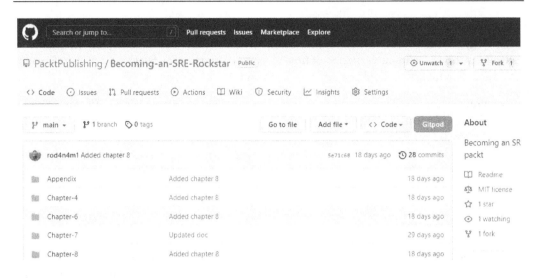

Figure 11.2 – Gitpod embedded on the GitHub page

Note how quickly you are taken to a full development environment; this is because Gitpod has a pre-configured container sitting at the ready, with tools and software built in already, so it launches in seconds, ready to work:

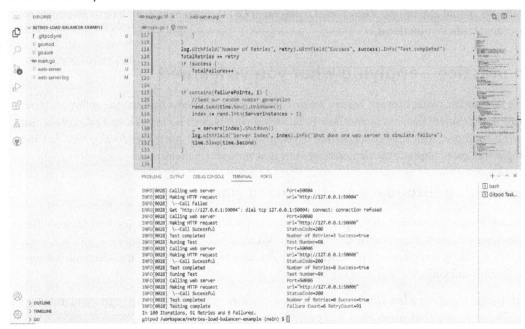

Figure 11.3 – The Gitpod workspace

To help show placement, we have colored this screenshot to show the separate regions of the Gitpod workspace. On the left, you'll find the standard navigation panel in green, and on the far left, in yellow, the navigation panel. At the top right, the file editor is at the top (in pink), while the terminal and output window tabs share the bottom area in blue.

The emulation source code

The emulation runs in a simple Golang application that is broken into two different parts. The first part starts up multiple web servers on different local ports, each with a health check. The second part continuously accesses those web servers, reporting back how many successful attempts are completed. The program allows you to specify the number of retries, starting with 0 or no retries when accessing the web servers, the number of web servers to start, and the number of web servers to randomly go offline during the emulation. You can see the command-line usage here:

```
-f int
      Number of Web Servers to Fail during test (default 1)
-i int
      Number of times to call the web servers (default 100)
-p int
      Base Port Number to start building servers on (default 50000)
-r int
      Number of Retries when calling to web servers fails (default 3)
-s int
      Number of Web Servers to Build. (default 10)
-v    Verbose Logging to Console (Info Level)              _
```

Figure 11.4 – Emulation command-line options

The verbose option will show what's happening behind the scenes, while the other options allow you to adjust the number of web servers, failures, retries, and the total number of calls to the web servers.

To build and run the code, simply download the required libraries as defined in the go.mod file, and then build the Go application:

```
go get && go build ./... && go test ./...
```

Running the emulation

When running the emulation, you should explore the use of retries and the number of web servers. Create a simple table of web servers, retries, and failures, and try different combinations of large and small numbers of each:

Servers	Retries	Failures	Total Retries	Total Failures

Figure 11.5 – Emulation scorecard

To run an emulation, insert the values as options in the command line – for example, this will run the example with 13 servers or workers, 3 retries, and 2 failures:

```
go run . -v -s 13 -r 3 -f 2
```

As we run the emulations, take note of the success rates. It should also be noted that the more web servers you have, the more it costs to run an application – be mindful of costs, and try to find the best number of retries across the fewest web servers.

Summary

Serverless technology and containers are here to stay. Their place in IT, especially when used in orchestrations, is amazingly powerful, giving us so much capability to provide even more reliability and resiliency in platforms.

Serverless tends to be a less intimidating technology to learn, but at high production levels – say, over a million calls a day – the level of complexity can grow. Even so, for nightly jobs, low volumes, and even compliance and testing, serverless environments can be extremely cost-effective places to do work.

On the other hand, containers and the many ways to orchestrate them are a subject area all their own. There are people who spend their entire careers inside of Kubernetes, running millions of containers in a single orchestration across multiple data centers around the world. You'll even find specialty divisions dedicated just to running these types of large-scale operations. However, remember, the concepts are the same; containers are simply self-contained workers, built to be reusable and easily tested and trusted as whole deployable units. This method of freezing code and systems is, by far, the most powerful piece of technology we have in the fight for reliability.

Speaking of reliability and resiliency, we will dive next into the world of testing and capacity planning. I've said many times that testing is, by far, the most proactive tool we have to build reliability. We will pair testing with capacity, where we will explore the limitations of our system's ability to do work.

12

Final Exam – Tests and Capacity Planning

If you want something to be reliable, there's nothing more necessary than testing it for reliability. Although it seems obvious to declare that, testing and tests are often poorly executed or wholly ignored. **Site reliability engineers** (**SREs**) must have a minimum knowledge of software testing during its development and build. Moreover, they need to incorporate certain techniques from the software testing framework into the monitoring stack.

Another critical aspect of systems is growth. Even when there are no extensive marketing campaigns, it's natural that users keep coming to consume services if the application is minimally compelling, which translates to a greater workload over time. SREs project a system's growth curve by applying the capacity planning discipline to ensure no saturation or disruption will happen.

In this chapter, we're going to cover the following main sections:

- Understanding types of testing
- Adopting **test-driven development** (**TDD**)
- Using test automation frameworks
- Staying ahead with capacity planning
- In practice – applying what you've learned

We will start this chapter by understanding the various software tests and why each has a distinct purpose. Then, we will discuss why TDD is a good practice and why SREs should be aware of it. We will check how to automate software testing by utilizing a framework so that it doesn't become *toil*. We will also cover how capacity planning helps us stay ahead in terms of the system's reliability. At the end, we will go through a simulation lab to learn how SREs do software load testing.

Technical requirements

At the end of this chapter, you'll find a testing simulation lab based on a **Node.js** application development. We want to assure you leave this chapter with valuable knowledge from it.

You will need the following for the lab:

- A laptop with access to the internet
- Node.js available on your laptop
- An account on a cloud service provider (we recommend Google Cloud Platform; you can create a free tier account here: `https://cloud.google.com/free`)
- kubectl and k6 installed on your laptop
- The gcloud CLI tool installed on your laptop (if you are using GCP)

You can find all files about this chapter on the public GitHub at `https://github.com/PacktPublishing/Becoming-a-Rockstar-SRE/blob/main/Chapter12`.

We start this chapter by examining which types of testing are essential for SREs.

Understanding types of testing

Suppose we consider testing a simple check against an expected behavior or value for a set of parameters; then, testing spans beyond software development. With careful consideration, you may see monitoring as a testing technique for production systems. SREs constantly test systems for reliability, from development to build and from deployment to running. The following figure displays the various types of testing:

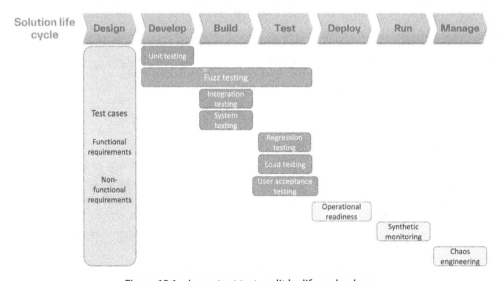

Figure 12.1 – Important tests split by life cycle phase

This diagram groups the critical tests by the solution life cycle phase, from development to management, passing through build, test, deploy and run. For that purpose, we divide this topic into five areas:

- Development tests
- Build tests
- Delivery tests
- Deployment tests
- Production tests

Let's explore those groups next.

Development tests

Andy Hunt, a programmer and author, once said, "*No one in the brief history of computing has ever written a piece of perfect software. It's unlikely that you'll be the first.*" This phrase is one of our favorite quotes in software development, and it reflects the core truth that we need to test software to identify bugs, which will definitely exist.

At this stage, we run **unit tests** to confirm a new functionality or feature has the desired outcome according to the design. Software engineers and developers write test cases and scripts to accomplish that. Assume we have a simple sum function:

```
function sumMe (factor1, factor2) {
    return factor1 + factor2;
}
```

It just returns the sum of the values passed as arguments. A possible unit test would be as follows:

```
if (sumMe(factor1, factor2) == expectedResultOf(factor1,
factor2)) {
    console.log("Test successful with ",factor1," and ",
factor2);
} else {
    console.error("Test failed with " ,factor1," and ",
factor2);
}
```

If you are new to software testing, you will reach the realization that we use code to test code. A direct consequence is that we automate unit tests by running test scripts that compare outcomes with expected results. This process eliminates many bugs that are not obvious or trivial, improving the software's quality and reliability.

Besides unit testing, a new type of testing has been gaining traction in the open source software community, called fuzz testing. It's the process of testing API endpoints with generated data from **fuzzers**. For the same function shown previously, a fuzzer would create data not expected by the JavaScript function to exploit any vulnerabilities. For instance, the following data points can break the code:

```
factors = [
    "a",
    "aaaaaaaaaaaaaa",
    "aaaa//",
    "hello/n",
    ...
];
fuzzer = {
    factor1: factors[random(factors.length)],
    factor2: factors[random(factors.length)]
}
test(sumMe(fuzzer.factor1, fuzzer.factor2))
```

Fuzz testing is especially effective for detecting vulnerabilities in the code, and it can catch memory leaks, data races, infinite loops, and deadlocks, among other code problems.

Of course, those tests are necessary, but not enough. We will explore software builds next.

Build tests

After we are done with **unit testing**, testing the functionality of individual pieces of code, and **fuzz testing**, trying to cause random problems with our code, it's time to put those units to work together as a single application or system. The goal is to verify whether any code changes have inserted a problem that has broken the build.

We run **integration and system tests** in this phase by utilizing specialized software testing suites. Integration testing exercises software grouped modules with pre-determined datasets and tests whether specific modules can work together after code additions or changes. It assumes that each module was unit-tested previously. Afterward, system testing works out the whole integrated application or software, checking the system behavior based on the specific functional and non-functional requirements. Also, system testing covers whether the intercommunication among different software modules and artifacts works correctly.

Next, we check the last tests before delivering the software.

Delivery tests

The system or software pack must pass one last set of tests before being considered for release. The first type is load testing, where we stress the application with high service demand levels to understand its capacities and determine the saturation and disruption points. The saturation point indicates how many users can make requests to the system before it becomes unresponsive or performs inadequately. The disruption point is the limit on requests at which the system crashes and has a non-recoverable situation without intervention. This test helps us comprehend the system's capacity and plan better for infrastructure requisites. As a counterpoint, the test environment needs production-equivalent IT resources to run a load test.

> **Important note**
> Often, companies put aside load tests as they consume production-level resources, which have high costs. However, planning for system capacity and growth is impossible without load-testing results.

We also employ regression tests in this phase. Regression testing runs unit, integration, and system tests with every software module to ensure a code change doesn't affect other areas of the software. Remember, we don't do unit tests on modules that didn't change in the typical path. This type of testing is a method from the **eXtreme Program** (**XP**) **Agile** methodology, which has an expensive toll. Executing a regression test every time something changes in the code may seem excessive, but it has proven valuable for interdependent systems.

The last test in this section is user acceptance testing – or just acceptance testing. Here, we use genuine or synthetic users to thoroughly test the software on the transaction level. What happens in this testing is we check the usability and reliability of the application services and features. This includes system monitoring and observability improvement through capturing and analyzing **metrics, events, logs, and traces** (**MELT**).

Automation is vital for full DevOps adoption. SREs must aid DevOps engineers in adding reliability tests to the testing stack. Next, we will dig into tests that are not related to software development.

Deployment tests

From an SRE perspective, a release candidate must go through the last test before being deployed to production. It's part of the **operational readiness review** (**ORR**) principle, where an application release and its underlying infrastructure have to meet a minimum set of conditions. Although we have extensively proven and tested the software version, we still need to verify its ease of maintenance and management. In other words, we inspect for two other architectural qualities: **maintainability** and **manageability**.

The SREs lead the ORR process by checking critical aspects of the application version candidate and its infrastructure. Initially, this process is just an agreed-upon checklist between the operations and development teams. This inventory includes available runbooks to operate and diagnose the system and its documented monitoring MELT – the defined service level indicators and objectives.

If you are thinking that the ORR checklist seems like hard work, you're right! SREs try to automate this checklist to avoid it becoming a bottleneck in a DevOps continuous deployment.

Production tests

We again apply real-user and synthetic monitoring to keep the production system under surveillance from a user point of view. Monitoring, in general, is a type of non-disruptive testing. The key difference in this phase is that we are not interested only in whether an element of the system fails the test but also in knowing how soon the component will cause the whole system to fail. For that reason, SREs implement **application performance monitoring (APM)** and **AIOps** for continuous testing and the analysis of a system's performance, following the observability principle.

We also have disruptive testing for systems that have excellent resiliency and availability. This is called chaos engineering, which uses bots to break the system parts to check whether it recovers or not and at what velocity. In *Chapter 16, Chaos Injector – Advanced Systems Stability*, we will cover chaos engineering.

Now that you know all about testing, from development to production, let's investigate how we accomplish that in software development in sequence.

Adopting TDD

Testing is so critical that organizations eventually devised a programming style and mindset called **TDD**. With this approach, we write the test case and run the test on the empty code. The code will fail against the unit testing, and we will start to improve it until it passes the test. This model may seem futile initially, but training a developer's mind to conceive of code as needing to pass a test has many benefits. For starters, it ensures that software testing is available and valuable right up front. With TDD, we only develop tested features that augment code reliability. SREs can and should adopt TDD for their coding tasks.

But how does TDD work? We can divide this topic into two sections:

- Unit testing the hard way
- Unit testing with a framework

We will first check how to accomplish TDD without using a tool.

Unit testing the hard way

We will use **Node.js** for this chapter, but you can find plenty of examples in other languages, such as **Python** and **Golang**. As mentioned, we start with the unit test. Say we want to develop an average calculation function. One way of testing it is the following:

```
function testMyAverage() {
    return myAverage([5, 10, 15]) === 10;
}
```

We know the simple average of 5, 10, and 15 is 10, but since the myAverage function doesn't exist, it will fail when we run the test script. That's the actual TDD!

We start developing the average function to pass this test. A possible solution would be as follows:

```
function myAverage(values) {
    const sum = values.reduce((acc, cur) => acc + cur, 0);
    return sum / values.length;
}
```

Unfortunately, this is not scalable, as we need to write more test cases. We may also want to add a fuzz test to check for vulnerabilities in this function. Next, we will study how to do TDD efficiently.

Unit testing with a framework

One of the most known test frameworks is jest. With it, you can assert functions with a single line of code:

```
test('Testing myAverage function - iteration: '+i, () => {
    const data = myAverage(testData[i].factors);
    expect(data).toEqual(testData[i].result);
})
```

jest allows complex assertions with data structures, which alleviates developers from worrying about writing test cases. In this example, we have the data sample in item.factors and the expected result in item.result.

Even more interestingly, you can create a loop to test multiple data points:

```
testData.forEach((item,index) => {
    test('Testing myAverage function - iteration: '+index, ()
=> {
                const data = myAverage(item.factors);
```

```
                              expect(data).toEqual(item.result);
            });
     });
```

Unit testing is the starting point for more reliable coding, while TDD with a framework is a lightweight approach to enforce testing from the beginning. We will expand on automation frameworks for testing next.

Using test automation frameworks

Good software testing doesn't necessarily imply a high-quality and reliable system, but no testing will most likely result in a flawed application with bugs. Despite that, companies that adopt DevOps and CI/CD pipelines have a clear requirement for automated tests, and they cannot afford long manual testing cycles that clog up their pipeline (and delay the release process).

Test automation frameworks aid developers and testers with good practices, rules, and guidelines that they follow pragmatically to achieve automated testing in software development. These frameworks establish how to write automation test scripts and manage test datasets. Within software engineering, we have the following types of test automation frameworks:

- Linear scripting

- Modular testing

- Library architecture scripting

- Data-driven testing

- Keyword-driven testing

- Hybrid testing

- **Behavior-driven development (BDD)** testing

Each type of test automation framework has pros and cons, and the discussion about which is better for a particular application development goes beyond the scope of this book. SREs must make use of the same framework as the testers. We will employ **Selenium** as a data-driven testing automation framework to illustrate how a test automation framework is implemented.

In the following diagram, the test script reads the test data from an external source and then executes planned operations in the target application with retrieved data points. Later, it compares the actual results of the application with the expected results stored with the test dataset:

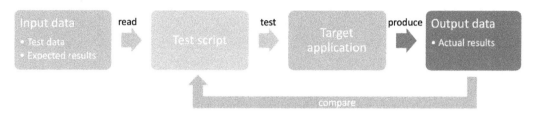

Figure 12.2 – Selenium data-driven testing automation

Selenium supports other test automation frameworks and spans multiple programming languages, hence why many companies adopt it for various types of testing, which includes user acceptance tests. **Selenium WebDriver** is a browser automation library broadly used for testing web applications. We can use it for any task that requires automating interactions with the browser. The following source code exemplifies how we apply WebDriver to test the Google search feature:

```
const WebDriver = require('selenium-webdriver');
const seachText = "rockstar SRE";
const regexStr = new RegExp(seachText,"i");
async function runUATest() {
  let driver = new WebDriver.Builder()
    .forBrowser('firefox')
    .build();
  // ncr stands for no country redirect
  await driver.get('https://www.google.com/ncr');
  const inputField = await driver.findElement(WebDriver.
By.name('q'));
  await inputField.sendKeys(seachText, WebDriver.Key.ENTER);
  try {
    await driver.wait(WebDriver.until.titleMatches(regexStr),
5000);
    driver.getTitle().then((title) => {
      if (title === `${seachText} - Google Search`) {
        console.log("User acceptance test passed");
      } else {
        console.error("User acceptance test failed");
      }
    });
  } catch (e) {
    await inputField.submit();
```

```
  }
  await driver.quit();
}
runUATest();
```

This code instantiates a `WebDriver` object that mimics a functioning browser – in this case, a **Firefox** browser. This `WebDriver` instance opens the Google website, inserts text in the search field, and then hits the *Enter* key. Exactly as a human being would interact with a search engine, it waits until the results are rendered and checks the pages' titles to validate that the search worked.

Now, we will investigate the nuances of planning a system's capacity to avoid crises.

Staying ahead with capacity planning

Capacity planning was vital to upholding operations not long ago, and it's still critical for traditional on-premises IT. In a world of hybrid multi-cloud computing, swift and precise predictions on the growth of the computing and networking capacity are still mandatory when hardware acquisition takes weeks and costs a lot. In the cloud computing era, and with the virtualization of many hardware functions, the importance of capacity planning is shifting from business survival to cost optimization.

Even with per-use payment and virtually unlimited resources available to businesses using hyperscalers, there is a growing concern in terms of cloud infrastructure costs and pricing models. Having a good idea of how the application workload expands (and how fast) is a standard way of planning for new virtual machines, clusters, or serverless environments. Reserving cloud services beforehand is suitable for staying prepared but also optimizes costs, as most cloud platforms offer discounts in these cases.

Capacity planning for either cloud-native or traditional on-premises IT environments needs at least three elements to work well:

- Load test data
- A capacity curve
- A demand curve

Let's start with performance testing.

Load test data

A critical part of software testing is determining whether a solution can serve a certain number of users. In other words, we want to find out how much load the application and infrastructure can handle in its current conditions. However, load and performance tests have another objective: determining the requirements for IT resources per workload level.

SREs implement load testing tools such as **k6** to run a sequence of performance tests for an application in a pre-production environment. They do so while measuring telemetry on CPU, memory, and network consumption. This report will indicate, for a certain number of virtual users, how many resources are required to maintain all services under nominal parameters. For illustration, say that an application spans 10 pods with 200m vCPUs and 100 MB of memory each when serving 100 users. An initial ratio for this example would be 20m CPUs and 10 MB per user. This theoretical proportion uses the requested resources as the basis for calculation, and we need to measure each pod's utilization if we want to have the actual ratio. We utilize the same thought process for network bandwidth and storage resources again.

With enough measurements, we can plot a capacity curve; we will see the details of that next.

The capacity curve

After measuring the consumed resources for the load demand, we can plot a scattered chart with the data points. We assume the load, say, the number of active sessions or users, is an independent variable for now, and it's on the *y* axis. Meanwhile, the monitored resource unit, let's take the number of vCPUs, is the dependent variable and the *x* axis of the chart. Using regression, a mathematical method to calculate the equation that expresses the relationship between two variables, we get the capacity curve for vCPUs. In the following diagram, we applied a linear regression method available in spreadsheets, and the resulting equation is $Y = 11.215X + 8.4091$:

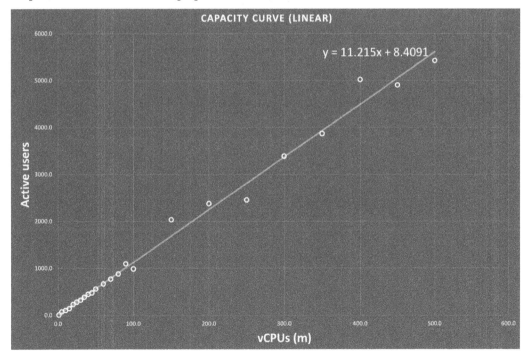

Figure 12.3 – Scattered chart with linear regression

If we replace the X variable per number of users in this formula, it will throw us the estimated quantity of **millicore** vCPUs (the Y variable) necessary for this demand level.

Usually, it's uncommon to see any resource following a linear capacity model, but it's the most straightforward model to learn. Nevertheless, spreadsheets such as **Excel** offer exponential, logarithmic, polynomial, power, and moving average regressions, and one of those models will fit the data points better. Also, it's possible to apply **artificial intelligence** (**AI**) to model the relationship between demand and resources, called workload profiling.

Unfortunately, the demand is not an independent variable as assumed initially, as it varies over time. Let's see how to handle this next.

The demand curve

Since the load or demand varies over time, we need to comprehend how it changes. Do we have repetitive peak loads? After the company launches a marketing campaign, do we expect more active sessions? As with the capacity curve, we use regression or AI to model how the workload grows and shrinks over time. SREs monitor the traffic (golden signal) for all user-interfacing applications to analyze the load behavior over time.

By understanding the estimated resources required for a projected peak load, the SREs can take advantage of better prices commonly available on cloud-reserved resources, or they can devise automated scripts for scaling resources up and down based on the capacity plan.

Let's dive into a practice lab to consolidate the learning from this chapter next.

In practice – applying what you've learned

As usual, we will close this chapter with hands-on experience with software tests in this testing simulation lab.

You will need pre requisite knowledge to appreciate this lab, as follows:

- Familiarity with JavaScript and Node.js
- Basic understanding of containers and **Kubernetes**

We divide this practical lab into three sections:

- Lab architecture
- Lab contents
- Lab instructions

Let's begin with understanding the design for this lab first.

Lab architecture

This lab employs the **k6** tool from Grafana Labs, an open source load and performance testing tool for applications, on a containerized **Node.js** application running inside a **Google Kubernetes Engine (GKE)** cluster.

It would be best to have the kubectl and k6 command-line tools installed on your laptop and the kubeconfig file (~/.kube/config) set for the **GKE** cluster under your GCP account (assuming you're going to use Google Cloud).

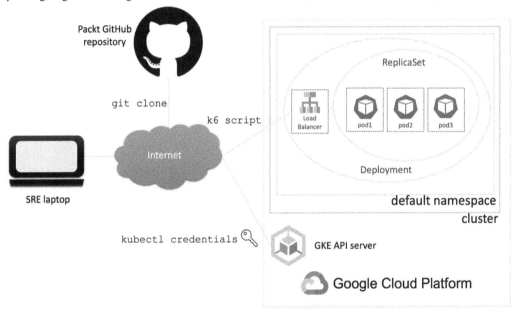

Figure 12.4 – Testing simulation lab diagram

In the lab, we will deploy a simple **Node.js** app using the kubectl command and run a load testing script with k6 to test this Kubernetes application. The application manifests have a **LoadBalancer** service, a **NodePort** service, and a **Deployment** with three replicas.

Lab contents

You can clone the GitHub repository by entering the following command in your terminal:

```
$ git clone git@github.com:PacktPublishing/Becoming-a-Rockstar-
SRE.git
```

Within this repository, under the Chapter12 folder, there is just one subfolder called k6. Also, there's a quick setup procedure called testing-simulation-lab.md in the same place. This procedure details the installation of the **command-line interface** (**CLI**) tools.

To access the **Grafana k6** content, go to this directory:

```
$ cd Becoming-a-Rockstar-SRE/Chapter12/k6
```

Inside this directory, you will find the simple-test.js test script file. It contains the **k6** instructions to run a simple HTTP test with one virtual user:

```
import http from 'k6/http';
import { sleep } from 'k6';

export const options = {
  vus: 1,
  duration: '30s',
};

export default function () {
  http.get(`http://${__ENV.GKE_ALB_IP}:${__ENV.GKE_ALB_PORT}`);
  sleep(1);
}
```

The **k6** testing framework operates using the **JavaScript** language for its test scripts. The first block contains the import statement, which loads the necessary built-in functions. Next, we have options including the number of **virtual users** (**VUs**) and the test duration. The last part holds the default test function definition. In this case, it makes an HTTP request to a URL based on environmental variables (the GKE cluster's **LoadBalancer** IP address and port).

We have a second test script file called load-test.js in the same directory for load-testing purposes. This file contains the k6 instructions to execute a staged load test against the same application:

```
import http from 'k6/http';
import { sleep } from 'k6';

export const options = {
  stages: [
    { duration: '5m', target: 100 }, // simulate ramp-up of
traffic from 1 to 100 users over 5 minutes.
    { duration: '10m', target: 100 }, // stay at 100 users for
```

```
10 minutes
    { duration: '5m', target: 0 }, // ramp-down to 0 users
  ],
  thresholds: {
    'http_req_duration': ['p(99)<1500'], // 99% of requests
must complete below 1.5s
  }
};


export default function () {
    http.get(`http://${__ENV.GKE_ALB_IP}:${__ENV.GKE_ALB_
PORT}`);
    sleep(1);
}
```

Like the previous example, the load test script has three parts. You may notice the second block has two arrays, one called `stages` and another called `thresholds`. Since a load test is a staged testing approach, `stages` objects possess the durations and target the VUs for each step.

The test script provided will simulate a ramp-up going up to 100 VUs within 5 minutes. In the next 10 minutes, it will hold 100 VUs hammering the application continuously. Then, in the last 5 minutes, it will emulate a ramp-down from 100 to 0 VUs.

The `thresholds` object holds metric names and their percentiles for the expected application performance. In this example, it expects that 99% of all requests made to the application have an HTTP duration shorter than 1,500 ms.

The last block is no different from the first script; it defines the default test function, which makes a simple HTTP request to an URL.

Next, we verify how to run this lab from your laptop.

Lab instructions

After you configure the **GKE** cluster with the information from your **GCP** project ID, it's time to run this lab.

We assume that you have installed and configured `kubectl` in the following procedure and that you have the `k6` tool installed on your machine:

1. Go to the lab directory and then the `k6` subdirectory. Use the following commands for that:

    ```
    $ cd Chapter-12; cd k6
    ```

2. Deploy the application to the GKE cluster using `kubectl`:

```
$ kubectl apply -f ../node/
```

3. Please wait for the **LoadBalancer** service to be assigned its external IP address. You can check the Kubernetes service with `kubectl`:

```
$ kubectl get svc node-api-rod-lb
```

You should see a piece of similar information to the one displayed here:

```
NAME                    TYPE              CLUSTER-IP     EXTER-
NAL-IP      PORT(S)                 AGE
node-api-rod-lb    LoadBal-
ancer    10.40.0.166    34.176.178.14    60000:32165/
TCP    84m
```

4. Configure the environmental variables. You can run the following script:

```
$ source process.env
```

5. Run the simple test script first. Use the following command:

```
$ k6 run simple-test.js
```

Here's an example of the output if you're running locally:

```
running (0m31.0s), 0/1 VUs, 29 complete and 0 interrupted iterations
default ✓ [===================================] 1 VUs  30s

    data_received...................: 7.2 kB 233 B/s
    data_sent.......................: 2.5 kB 80 B/s
    http_req_blocked................: avg=2.25ms  min=3µs    med=5µs    max=65.17ms p(90)=7.2µs  p(95)=15.19µs
    http_req_connecting.............: avg=2.23ms  min=0s     med=0s     max=64.73ms p(90)=0s     p(95)=0s
    http_req_duration...............: avg=65.89ms min=61.19ms med=67.69ms max=71.39ms p(90)=68.64ms p(95)=69.57ms
      { expected_response:true }....: avg=65.89ms min=61.19ms med=67.69ms max=71.39ms p(90)=68.64ms p(95)=69.57ms
    http_req_failed.................: 0.00%  ✓ 0        ✗ 29
    http_req_receiving..............: avg=66.58µs min=36µs   med=63µs   max=121µs   p(90)=90µs   p(95)=103.39µs
    http_req_sending................: avg=29.68µs min=12µs   med=22µs   max=206µs   p(90)=35µs   p(95)=35.6µs
    http_req_tls_handshaking........: avg=0s      min=0s     med=0s     max=0s      p(90)=0s     p(95)=0s
    http_req_waiting................: avg=65.79ms min=61.13ms med=67.59ms max=71.06ms p(90)=68.55ms p(95)=69.46ms
    http_reqs.......................: 29     0.935538/s
    iteration_duration..............: avg=1.06s   min=1.06s  med=1.06s  max=1.13s   p(90)=1.06s  p(95)=1.07s
    iterations......................: 29     0.935538/s
    vus.............................: 1      min=1      max=1
    vus_max.........................: 1      min=1      max=1
```

Figure 12.5 – Example of output for the simple-test.js execution

Notice there were 29 complete iterations, almost 1 per second.

6. Run the load test script next. Use the following command:

```
$ k6 run load-test.js
```

Here's an example of the output after 20 minutes:

```
running (20m00.9s), 000/100 VUs, 84881 complete and 0 interrupted iterations
default ✓ [======================================] 000/100 VUs  20m0s

     data_received...................: 21 MB   18 kB/s
     data_sent.......................: 7.2 MB  6.0 kB/s
     http_req_blocked................: avg=79.63µs min=1µs    med=5µs   max=73.36ms  p(90)=7µs     p(95)=8µs
     http_req_connecting.............: avg=74.11µs min=0s     med=0s    max=73.25ms  p(90)=0s      p(95)=0s
   ✓ http_req_duration...............: avg=62.05ms min=56.63ms med=61.55ms max=368.53ms p(90)=64.43ms p(95)=64.92ms
       { expected_response:true }...: avg=62.05ms min=56.63ms med=61.55ms max=368.53ms p(90)=64.43ms p(95)=64.92ms
     http_req_failed.................: 0.00%   ✓ 0          ✗ 84881
     http_req_receiving..............: avg=65.41µs min=14µs   med=63µs  max=1.21ms   p(90)=95µs    p(95)=101µs
     http_req_sending................: avg=23.09µs min=5µs    med=21µs  max=695µs    p(90)=36µs    p(95)=39µs
     http_req_tls_handshaking........: avg=0s      min=0s     med=0s    max=0s       p(90)=0s      p(95)=0s
     http_req_waiting................: avg=61.96ms min=56.59ms med=61.46ms max=368.47ms p(90)=64.33ms p(95)=64.83ms
     http_reqs.......................: 84881   70.681616/s
     iteration_duration..............: avg=1.06s   min=1.05s  med=1.06s max=1.36s    p(90)=1.06s   p(95)=1.06s
     iterations......................: 84881   70.681616/s
     vus.............................: 1       min=1      max=100
     vus_max.........................: 100     min=100    max=100
```

Figure 12.6 – Example of output for the load-test.js execution

The k6 tool reports after completing load testing. This test is an extended cycle test with more than 84,000 iterations. There's a green checkmark on the `http_req_duration` row because we have configured it as a threshold for this test. Since the application answered all requests with a maximum of 368.53 ms, it passed the test, as the target was within 1,500 ms.

7. You have done it! Congratulations!

You can run this lab with other hyperscalers, such as AWS and Azure. Remember that you can run the **k6** test script from within a Kubernetes cluster or the k6 cloud environment if you need more power for your load testing. SREs are synonymous with testing, whether applying software tests, reviewing operational readiness, or monitoring with test automation frameworks.

Summary

Testing is the only way to have a reliable application and viable operations. We illustrated the importance of this area of work and the prevalent practices within it. We also examined how capacity planning is intrinsically related to testing. Even though it's an entirely distinct area of work, we clubbed capacity planning together in the same chapter because of this relationship. At the end of this chapter, we demonstrated a lab to help you consolidate this chapter's content with practical knowledge.

By now, you should be able to articulate the various software tests in the different stages of the solution/ software development life cycle. Test automation frameworks are essential for DevOps and the existence of a CI/CD pipeline. We came to understand how operational readiness reviews, monitoring, and chaos engineering are types of testing, and how a reasonable capacity plan helps an organization stay ahead in terms of preparedness and cost.

In the next chapter, you will master outages with an SRE's mindset. We will talk about runbooks and low-noise notifications.

Further reading

- To learn more about TDD, please visit the *freeCodeCamp* article here: `https://www.freecodecamp.org/news/test-driven-development-tutorial-how-to-test-javascript-and-reactjs-app/`

- To learn more about creating a test automation environment with Selenium for cross-browser testing, we recommend the following website: `https://developer.mozilla.org/docs/Learn/Tools_and_testing/Cross_browser_testing/Your_own_automation_environment`

- For a load and performance testing tool, please use this link: `https://k6.io/docs/`

- You can read about ITIL Capacity Management from the *IT Process* wiki at this link: `https://wiki.en.it-processmaps.com/index.php/Capacity_Management`

Part 4 -
Mastering the Outage Moments

Working an outage should be akin to a chef running a well-oiled kitchen with readily able staff and well-defined tasks assigned for different purposes. First, we discuss how to run an outage and the unique training opportunities it brings with it. Then, we review how to best prepare for outages, and finally, when we're in the thick of it, how to leverage the different types of people we often see in these situations. In closing, we'll discuss external messaging to executives and customers – and why understanding perception is important.

The following chapters will be covered in this section:

- *Chapter 13, First Thing – Runbooks and Low Noise Outage Notifications*
- *Chapter 14, Rapid Response – Outage Management Techniques*
- *Chapter 15, Postmortem Candor – Long-Term Resolution*

First Thing – Runbooks and Low Noise Outage Notifications

Institutional knowledge is the experience and understanding that staff have about a company. From the long-standing engineers who wrote the code that runs the day-to-day business, to network engineers familiar with the topology of the data center, and even customer service representatives, who know how to leverage internal company processes to resolve customer issues. This employee understanding takes years to build in employees and can be difficult – if not impossible – to replace. Runbooks are the embodiment of institutional knowledge about applications and systems in a tangible document to aid in troubleshooting and resolving issues.

Runbooks may embody the knowledge needed to resolve issues, but we must combine them with alerts that tell us when an outage occurs. When I think of alerts, I'm always reminded of the boy who cried wolf – the story of a boy who is bored at night tending sheep and cries wolf night after night. After multiple nights of being woken from bed to find the little boy was only lonely and wanted attention by crying wolf, the village stopped responding to his wolf calls. The night the wolf really did show up, he lost his sheep because nobody believed him anymore.

This alert de-sensitivity is a very common issue in technology, and especially for Site Reliability Engineers. When alerts continue to just fire and fire, they often get ignored – effectively rendering your alert worthless. This chapter will focus on both runbooks and alerting notifications. We will discuss how dashboards are part of both the problem and solution in SRE.

We are going to cover the following main topics:

- What makes a good runbook – the basics
- Beyond the runbook – code and comments
- What's in a good dashboard?
- The basics of priority levels

- Alerting with priority levels
- In practice – applying what you've learned

Technical requirements

For this chapter, you will need an understanding of basic system architectures, as presented earlier in the book, and an understanding of API and database connectivity to applications. In addition, you'll need knowledge of basic logging, alerting, and dashboards.

What makes a good runbook – the basics

Okay, I'll admit it – I'm personally not a fan of runbooks. I just haven't found a better tool to make the knowledge I need available to me instantaneously during an outage. For me, it's a struggle between wanting to be the type of engineer who can walk into an outage and in minutes resolve the issue and knowing the truth – I'm not always as good as I think I am or want to be. And that is the most honest reason why we need runbooks.

One of the strong characteristics most rockstar engineers have is the ability to solve problems and find a path to resolution quickly – whether in an obscure compiler error or the highest priority outages. We leverage this power to quickly determine cause and remediations, but imagine if we had the article with the answer bookmarked – how much time would that save? That is the strength of the runbook – it's a singular answer to understanding not only an application or process but also the bookmarks for dashboards, log queries, technical documentation, and more – all ready to be clicked on. Runbooks are about saving time.

Runbooks exist in all different formats, from Word documents to Git repositories, in Confluence, and even on printed paper. What goes into a runbook? And lastly, how do we ensure it's up to date? Let's explore these questions together.

Runbooks as living documents

I've seen runbooks in all sorts of formats in all sorts of collaboration systems. The best, though, are those runbooks that are forged as a shared living document. This living document should allow your updates to be readily available to all once saved. And most importantly, the latest revision of the document must always be shown and made available.

Why living documents? There is a simple answer. If your runbook exists in a format such as a Word document that is emailed to an entire department, then the moment a revision is needed, say, updating the phone number of the lead technician, a new version must be released and emailed out. So, what's wrong with this? As new versions are released, it requires everyone to save that file to their computer, or hunt for the latest version when an issue occurs.

When we compare this manually updated and distributed runbook file method to an online living document, we provide a common, bookmarkable link to the runbook. And every time someone comes to visit the runbook, they always get the most up-to-date information.

Copying versus linking documentation

As we start exploring the use of living documents in runbooks, it can be easy to just copy and paste information from other sources into a runbook. And while this may be okay for basic information such as a description of the application, if a team has established its own source of documentation and information for an application, we want to avoid duplicating its effort by copying and pasting its information. Instead, runbooks should always use a link to documentation when possible. This allows teams to continue updating their own documentation and allows that information to be linked to the runbook for fast retrieval.

One item to remember when making links to documentation is that the person reviewing the runbook must have access to the links you use. So, for example, if you use a link to a Git repository, you want to ensure your entire runbook audience has the appropriate permissions to view that repository.

Understanding the runbook audience knowledge level

When we write runbooks, we must take into account the skill set of the audience who will be using the runbook. If your runbook is too simple for the highly technical audience that consumes it, or the runbook is too complex for entry-level engineers assigned to remediate issues, your audience may choose to discard the runbook. That being said, do not be afraid to prefix certain sections with the technical level required, for example, "This section was designed for Node.js/JavaScript developers."

It should also be noted that new engineers, especially in DevOps and SRE fields, will, regardless of their skill set, often have gaps in capabilities until they learn the systems and tools specific to companies. This timeframe is often quoted as 6 months.

Runbook audience permissions

We also should verify that the runbook audience has the correct permissions to take any outlined actions or to view system outputs and monitoring. If elevated permissions are required and not given to engineers, ensure you have an escalation plan in place as well. This escalation plan may state to contact a senior engineer or the **Network Operations Center** (**NOC**).

Another option is to have a mechanism that grants the runbook audience temporary elevated permissions just for the incident they are working on. This can help satisfy security requirements by only allowing access when needed and placing controls around both the elevation of privileges and the systems they apply to.

What do you put into a runbook anyway?

That's the question, isn't it? There are four basic elements I start off with in my runbooks:

- A description and documentation about the application, including diagrams
- Contact information for key subject matter experts, escalation points, and vendors
- Links to assets such as source code, log queries, alerts, and dashboards
- Associated third-party assets, including databases, APIs, and other vendors

These four subjects can become quite complex. The key is this document should direct any engineer – junior, senior, or rockstar – to be able to pick it up and immediately understand how the application works – its dependencies, sources, and destinations for data, and what observability is available for it. In short, everything you would teach a junior developer diving into an application for the first time.

Personally, as a staff or rockstar-level engineer myself, I think of runbook content as everything I'd ever want to know about an application, should I be asked to troubleshoot it with no prior knowledge or experience with it.

Runbooks are not everything. Next, we will look into how code and additional developer documentation can provide even more insight into an application.

Beyond the runbook – code and comments

Regardless of the amount of documentation an application has, the only true source to determine how an application works is its source code! The problem is that in today's world of coding, even source code can be tricky to read at times. And while developers do add code comments, aka notes embedded in the code, remember they are not always correct or up to date. I tend to trust comments more than any other type of documentation though.

Quickly understanding source code

When analyzing source code, you've just been handed – and yes, you'll have one of those outages at some time – there are a few tricks I've used over the years. But it always starts with finding the entry point where the code starts. This could be a function in the case of serverless functions, or simply the `main()` function inside `main.c`.

Most application frameworks these days also allow users to run code prior to the entry point outside of this entry point. From global variables to creating new instances of classes, and even running bits of code – understanding these prerequisites of a development language will give you insight into items that are often *hidden* in ways not always easily found.

You'll also find many application developers follow the same directory structure and wording, which can assist you in diving into the code you're looking for. In addition, filenames can be just as telling. In essence, when you review code, you'll often find a familiar feel to multiple applications inside an organization.

Searching source code for your needle in a haystack

When I start looking through source code, one of my favorite tricks is to look at the global variables and settings being brought into the application. Items such as database connection strings and API URLs can quickly be translated into their associated `settings` variable. Then, you simply search the code for this `settings` variable – and you'll be led directly to where an API or database may be used. This search technique may have to be done more than once. For instance, a database connection string leads to the associated `settings` variable, which leads you to the function that connects the database. Now you search for the `connect` function to find where the database may be used in code.

And of course, as covered in *Chapter 5, Resolution Path – Master Troubleshooting*, searching code by the errors shown in logs is an amazing way to pinpoint the code causing problems.

Commenting for understanding

When all else fails, I start line by line, adding my own comments to the code. This trick forces you to stop skimming and forces you to read the code line by line, process it, and then type out what it does. Trust me, when you've been skimming code and you can't find the issue, this amazing strategy can bring a new level of clarity.

Not only do we comment just for understanding but also, adding comments to the API request and response formats, and even actual request and response data samples, gives clarity into exactly what to expect. This technique is useful not only for troubleshooting but is also a great design aid – especially when your junior engineer is struggling to find the right JSON path to data!

Beyond the runbook and source code, the dashboards provide a quick insight into the current state of the system. At a single glance, dashboards can be quick ways to check whether your system is performing on par – especially when using a single pane of glass dashboard.

What's in a good dashboard?

That's a great question with a not-so-simple answer. Dashboard content is highly dependent upon the audience they are intended for. From **NOC** style dashboards simply designed to display whether services are up or not to complex dashboards used by developers to identify issues with the processing times of their application code – dashboards can serve a large number of purposes. I've even written dashboards for finance to quickly determine the impact when a system goes down – using log analysis.

Types of dashboards

As we've discussed, there is no limit to the types of dashboards we create as rockstar SREs. Your dashboards will be largely defined by the audience they are being created for. In discussing what should be on dashboards, take into account their technical skill set and understanding of the system, along with the depth they want to see. But in general, I think about three basic classes of dashboards.

Dashboards for executives and businesses

This type of dashboard is often for those people outside of the technical organization or in a leadership position within technology. We often look at this in terms of the big picture. Especially for businesses, you'll want to discuss with them exactly what the dashboard is to fill in and identify the key sources that can answer these questions.

In this type of dashboard, you must bring a high level of clarity and understanding – we discussed false positives before, especially in HTTP logs, and in many systems, externally exposed APIs can suffer uptime issues if these numbers include false positives such as bot traffic. This needs to be communicated and understood by the audience, or other methods of finding this data need to be explored.

This type of dashboard should be built with a higher amount of polish and, hopefully, fit on a single screen for easy consumption.

Dashboards for troubleshooting and diagnostics

When it comes to troubleshooting and diagnostics, we are not so concerned about what is pretty, but rather, how much data of a highly technical nature we can pack into the dashboard. The goal is visibility at a glance because the consumers of this type of information often need to see the raw data, including the false positives.

When we build these types of dashboards, we often want to include trending data, or graphs showing what the system has been doing over time – are error counts rising or falling? And comparisons of data with other data in time frames, for example, comparing web traffic against the same time last week.

These dashboards often lack polish and can even contain broken elements or hastily written notes and graph titles – and it's not uncommon to add things to a dashboard on the fly when in the middle of an outage. Make no mistake, these dashboards are some of the most powerful tools that technical teams can use to maintain systems.

Dashboards for developers and performance

Developers often want their own dashboards, especially for performance tuning. As code is built, we can introduce new functionality that causes latency or impacts third-party systems. In addition, architectural changes to environments bring possible performance bottlenecks visible during performance or load testing.

I enjoy seeing developers tinkering with their own dashboards. It's a sign of a true professional. And the use of observability systems to track development issues and progress gives an amazing new toolset to developers.

When developers understand your observability tools, this also gives you, as an SRE, an added edge in remediating production issues – as now, you have a whole team of engineers who can dig into observability data with you during an outage.

Single pane of glass dashboards

You won't find too many of these types of dashboards – because the purpose is to provide a singular view of a large part of a process, or the entire process. I liken these types of dashboards to a steam plant display, or a map showing points of interest where situations such as slowdowns or other issues may happen. These types of displays don't allow us to see every pipe and valve temperature and pressure – or how fast every car is going; instead, key points in the traffic or plant are drawn out to represent functionality.

In something as simple as a website, a single pane of glass dashboard may include total customer logins, items put in carts, orders started, and orders completed. In addition, we may add a single count of total errors on all the APIs and a list of APIs ordered by API, with the highest error count on top.

This display would provide us with a quick view, not of technical issues but of business problems – such as nobody completing an order because credit card processing is down. These types of dashboards show insight into functionality at an origination level and it's common when something on this page goes wrong to use our technical dashboards to pinpoint the issue.

So how are these different than executive dashboards? Executive dashboards are designed to answer business questions, and while they can be similar, they often have less dense data and lack the technical deep dive we add to single pane of glass dashboards.

NOC-style red and green

A **Network Operations Center** (**NOC**) is often thought about in the industry as a thing of the past, where people sat in large rooms and watched dashboards and supported outages as a group from a single location. With the expansion of incident management systems such as **PagerDuty** and **Opsgenie** paired with alerting, we can now simply reach out and bring expertise into an outage.

Old NOC-style dashboards had boxes, rows, and rows of boxes – each representing a different item, system, or server – that simply turned from green to yellow or red when an issue was detected. This simple but effective technique is still widely used in individual dashboard components.

There are very specific use cases for this antiquated type of dashboard, and the most popular still today is for desktop and first-level support engineers to have a quick view of what systems may be up or down.

Displaying trends

When we look at dashboards, it's often common to see data displayed over time referred to as **timespan** or **trend** graphs. This type of display of data provides us with a history of the measurements.

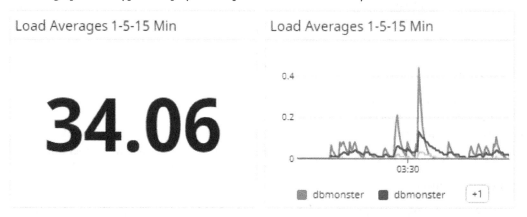

Figure 13.1 – Sample dashboard trend displays

When we look at data over time, it's very different than data presented as a single number. This display from a dashboard shows a singular view of the data, but notice the spikes when we look at the data as a trend.

Pattern detection

Trend graphs have the ability to show us patterns in a system. Given that many systems perform repetitive tasks ranging from data reporting to database cleanup and even backend systems work, it's often common to have tasks on a schedule.

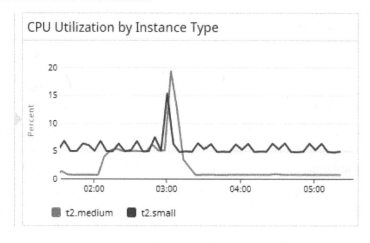

Figure 13.2 – Sample trend pattern graph

This graph shows that one of our instances is processing data on a schedule, as indicated by the repeated spikes.

When these scheduled tasks have an impact on a system, say a database table lock causing an impact on query times, seeing this pattern helps us identify issues specific to these tasks.

As a rockstar SRE, I prefer to use arbitrary numbers to schedule my tasks as well, say, 7 minutes past the hour. You'll find this technique helps distribute the load as other tasks take the more mundane on-the-hour approach. And if your timing is unique to your task, identifying the trend and its source becomes much easier.

Ramping up and ramping down

When you look at a dashboard that simply displays a count of errors in the past 15 minutes, you have no history as to whether the error count is rising or falling. Adding a trend graph allows insight into the source of the numbers. Why is this important? Without this, when you take corrective action, you may have to wait 15 minutes for a number to change in a box – whereas a trend graph shows the falling error count in minutes.

The same is also true as impacts start to take hold of the systems – by knowing the error count, latency, or other measurement is increasing, you have a better idea of the rate at which the system may be impacted.

Combined counts and trending

Having a count gives a sense of volume, and when you pair a count with a trend, this is my favorite type of display, only recently made available in observability tooling.

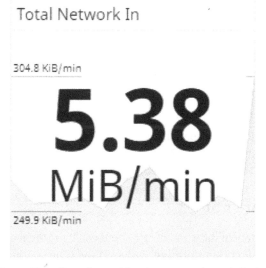

Figure 13.3 – Sample combined counts and trend display

As you can see, this simple display tells us both the volume and trend, allowing us to define the amount of impact, along with whether that impact is rising or falling.

Aggregates and breakdowns

Especially on single pane of glass and executive dashboards, we often create totals across multiple applications, servers, or even an entire platform. This allows a more simplistic view of the system, but when paired with a breakdown, it's an amazingly handy tool.

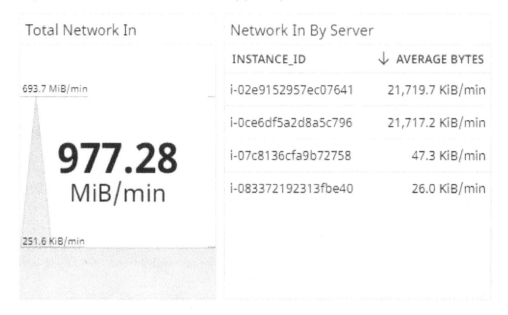

Figure 13.4 – Sample breakdown display

In addition to simply knowing the volume, in this example of network traffic across a number of servers, services, or even the entire network, now I can see what device is contributing the most (or least) to this aggregate number.

As a rockstar SRE, this is one of my favorite tricks. Sum all the errors from all the microservices into a single count so you know when something is wrong, then look at the list to identify the exact microservice having the issue. In addition, if you use a combined count and trend display, you can also see whether the issue is steady, increasing, or decreasing.

What dashboards are not

Dashboards are an amazing tool in the world of SRE – and you'll use them often. From seeing where the problem is to how a system is performing, and even the state of a problem – you'll come to love dashboards.

What dashboards are not is alerting. Dashboards can't wake you up at 3 a.m. when a server goes down or tell you when you're repeatedly running out of memory in containers. Alerting is the key to bringing engineers into issues – dashboards are the tools you use when you arrive. And how we send alerts is just as important, which is why we will talk about priority levels next.

Even though dashboards are not alerts, it is common practice for alerts to include links to at least a visualization of the metric causing the alert. You can even include links to specific dashboards.

The basics of priority levels

Why do we have priority levels? The answer – and the question is rarely answered correctly – is that priority-level systems retain engineers.

Executives will tell you priority levels define the severity of issues. Product managers will say they define the impact on the customer and business. Scrum masters will say they define the urgency of remediation. Engineers will tell you it's about whether they have to get up at 3 a.m.

While they are all right, priority levels truly drive two simple things: response effort and engineer retention.

Response effort

The level of effort should be equal to the priority of the issue – the higher the priority, the faster we expect the response to be (and the greater the number of people responding). This can follow along with remediation after the issue as well. A very noticeable outage of high priority will often have an impact on pre-planned work, adding new priorities to the workload – thus delaying work planned or in flight.

When responding to lower-priority issues, we should have fewer people responding, or the response might not occur until the next business day. Future work to remediate the issue completely may be scheduled weeks or even months after the event.

Engineer retention

There is no doubt that repeatedly calling people to work in the middle of the night, especially when they are expected to come in as normal the next day, is a perfect formula to cause attrition. Most of us in the industry understand, and even agree, that we may get called at any time – day or night – to help remediate issues.

Retention of staff will become an issue when we repeatedly call them in on emergencies that are not emergencies – or we artificially inflate the priority due to political or emotional connections to problems. Imagine getting called at 3 a.m. three days a week – each time for an alert that actually showed nothing was wrong. Imagine how tired and mad you would be! Now imagine a recruiter sends you a job description the next day on LinkedIn. How much more willing will your staff be to start thinking "Maybe I should start looking at other job opportunities."

To prevent staff attrition both inside and outside of your team, priority levels offer a guarantee that there is actually an emergency before you send an alert at 3 a.m. – and when it's not critical, the other alerts can wait until the next business day before alerting anyone. This type of priority-based alerting can be driven by **incident response systems**, with **PagerDuty** and **Opsgenie** being the two leading options.

Incident response systems and priority

Incident response systems allow us to manage incident communication in a whole new way. In the past, often, one person called others into emergencies, hoping that after 6 calls at 3 a.m., a lead engineer would wake up and take the call. Or, worse, phone numbers would not be available to those who needed them. It would become a game of whose phone number do I have, and who do I know who's most likely to take my call at 3 a.m.?

Instead, imagine you put teams on an on-call schedule that automatically rotates and even gives employees the ability to cover each other's shifts. Imagine that for a high-priority issue, employees would be contacted more aggressively and that this even allows setting up a secondary on-call engineer to be called when the primary engineer doesn't respond. The system could even reach out to senior management or entire teams if neither engineer responds at all. And finally, you could also add priority levels to this system, allowing only critical issues to alert your staff after hours, holding everything else till the next business day. This is only part of what incident response systems do.

Incident response systems and phone-based alerts

One of the most powerful features of today's incident response systems is the ability to manage different ways of communicating, via a smartphone. With the mobile apps available for these systems today, we can not only define different notification tones and visual alerting but also bypass a do not disturb setting to reach us.

When an alert comes to my phone, if it's of critical priority, the app will bypass any do not disturb or volume settings and ensure I receive the most critical alerts, while less critical alerts may be allowed to be silenced.

What is a priority one event?

We talk about priority levels typically in terms of P1 through P5, where P1 is the most critical of emergency issues, and P5 might be thought of as an alert that you just want to investigate over the next few days. The definition is easy enough, but defining priority, especially in relation to the number of errors, drops in traffic, or a number of other metrics or SLIs, can be tricky.

But a priority one event? That's simple. This priority level is often considered as a total outage, complete failure, high impact, or a total lack of functionality. This is the priority level of a website being completely offline, or no customers being able to log in.

Defining priority based on...

When we look at alerts, these often look at glimpses of time, sometimes as small as 5 minutes. Those timeframes are a trade-off – how long can something be abnormal before I call it an issue? That timeframe is how long before you'll know something is wrong. Too short and small glitches such as software deployments can cause alerts that disappear before a responder can even look. Too long and it can be an hour before anyone knows something is wrong.

We can define alerting levels based on a number of different criteria, but the most common are revenue, customer impact, and impact on specific functions. As we go through each, think of how these apply to your SLAs.

Revenue

Business is business. Engineers may not understand the value each customer transaction brings, but product and business partners often have razor-sharp metrics that track revenue based on a large number of criteria.

These criteria will also look different depending on the business model. A site that sells shoes, for example, will tend to be very simple to understand – if you can't take credit card payments, that's a critical issue. Other systems, such as those that monetize social interactions using advertisers, can be far more complex, but again, simple functionality such as not being able to get onto the website or not being able to log in, can have a clear impact – ads are not being shown to users.

Impacted customers

Another favorite model is the count of user impact. This focuses on the total number of users impacted. This is extremely effective when dealing with error counts and transaction failures – it allows us to remove the monetization and have a clearer picture.

Functionality

As we've seen in the two different parts of our custom hat company website, there are functions that are more critical to the revenue path of a business than others. Understanding the relevance of that functionality can quickly identify higher-priority issues. For example, if you can't log in, that's often critical.

The priority level of observability failures

One interesting part often left out of alerting and monitoring is observability itself. What happens when your observability system stops functioning? The simple answer is you're blind.

Like all tracking of reliability, observability of the systems providing your observability should not be excluded, and this question is highly dependent on the level of observability redundancy you have. If you have separate log and metric analysis, can you set up alerts in both? Can your vendors or other teams your application connects to provide alerting on a different observability system?

You should define your own rules. For me, if I do not have redundancy, I classify any observability outage as a P2, to be resolved at any time – day or night. If I have observability redundancy, I classify it as a P3.

You might question how we know when observability is down. Well, for that, we have to enable observability on our observability platform. One of my favorite tricks is to set up observability platforms to talk with one another. For example, set up an alert that you know will always fire every 5 minutes, which calls an API endpoint, which creates a log entry in your logging system – and in your logging system, create an alert that will fire when no data comes in for 20 minutes.

Forcing the priority – the rockstar way!

Many observability systems allow you to define priority within the alert itself or even to define different priorities for alerts versus warnings. I personally prefer to force priority when possible.

What do I mean by forcing priority? Simple: when I select the communication interface, be it an email address, or integration, I always build out those integrations with built-in priority levels – for example, `devops-P1@custom-hats.com`. Or in Datadog, you may call integration `@PagerDuty-DevOps-P1`. This forces the person setting up the alert to be extremely mindful of the priority level, instead of hoping they select the priority level in a web form. This rockstar-level integration trick is amazingly effective and has helped me ensure priority levels are thought about and correctly assigned.

Adjusting alerts

Alerts can't be talked about without the realization that your alerts will not always go as planned. Even the best of intentions in alerting can sometimes fail, and we must be prepared at any time to make adjustments to alerting.

My teams have always made alerting adjustments a priority. Why? Simple: if an alert is falsely waking people or even just distracting them from their tasks at hand when nothing is wrong, then the alert gets ignored and becomes noise.

Logs and alerting

In the world of alerting and metrics, it's often easy to overlook implementing alerting based on the contents of logged messages. We often find information in logs found nowhere else. Remembering these unique messages in our alerting should not be discounted.

We have to be particularly aware of specific platforms. For instance, **AWS Lambda** has an error metric that is a count of Lambda failures – that is, the Lambda application crashes. If the application uses a `Try Catch` statement to capture and log exceptions but the application exits gracefully, then the error is not counted in the AWS CloudWatch metric for a Lambda error count, rather, it is only found in CloudWatch Logs.

Pausing alerts

Finally, when we have alerting, there are always exceptions – times when teams may be allowed time to resolve issues. During these times, we typically silence the alarm. While this seems like a good idea on the surface, it can leave applications unprotected against other outages. In particular, alerts generated from logs, when silenced, can cause all error messages to be ignored. Instead, the rockstar SRE should know how to add an exception for a specific error message, so all other errors will not be ignored.

When possible, it's also preferential to add an expiration to the pausing of alerts. Some systems such as Datadog allow for alerts to be muted for a specific period of time. In other logging systems, you can compound the date and log exception, like this:

```
source=aws-lambda type=error AND NOT (message = "error
processing button click" AND date.now < 2023/01/01)
```

This example would still send alerts when there are errors, but not if the log error was the specific error `"error processing button click"` and the error happened before January 1, 2023. This added date allows the exception time to expire to remind the SRE team to check with the team on updates if the error is still happening. In addition, the exception timeout ensures that if the error message happens in the future, say in the middle of 2023, the error will cause an alert.

In practice – applying what you've learned

In our practice exercise, we are going to build out a definition of priority levels, build the runbook for the custom hat pricing API, and define a list of alerts and their associated priority levels. As part of this exercise, we will build this documentation in GitHub, which provides both a living document platform and keeps a history of changes over time.

Defining priority levels

When we define priority levels, first define the information you want both product and engineering teams to both understand and commit to. This is the list I frequently use:

- Priority level (P1–P5).

- A simple definition of what priority means – this should be clear to both engineering and product teams.

- Who will be notified – this should be not only a list of who gets notified but also how those notifications change based on priority level. For example, leadership may want immediate notification of a P1 or P2 event alongside the SRE and development teams.

- Time frame notifications will be sent – this should define when an alert is allowed to be sent, for example, a P4 may be limited to 9 a.m. to 5 p.m. on weekdays. This also establishes an expectation for when teams are to respond, for example, if a development or QA environment issue happens, do we expect a response at 3 a.m.?

- The level of effort the development team makes to resolve the issue – knowing about an issue is important, but having a commitment from engineers as to their response provides clarity in how resources are allocated to resolve issues.

Custom hat pricing API runbook

For our runbook, we want to collect the following information:

- **Description**: A simple functional description.

- **Contact info**: Who to contact for issues, including emergency response, engineering leads for development, and data. You should also include escalation contacts should additional assistance be required.

- **Vendors**: Vendors you are dependent on, including their vendor contact information, links to the products used, manuals, API libraries, and, if available, a link to the vendor's status page. With vendor information, ensure you add account numbers and other identifying information for agents who will need to assist you.

- **Data**: A list of any data sources, that is, databases, and so on, and connectivity information including encryption information, schemas, and overviews of stored procedures, views, and scheduled jobs if used.

- **APIs**: A list of APIs and their documentation, both for the application itself and applications you are connected to.

- **Alerts**: A list of the alerts, or a link to a search for all the alerts for this application. You may also want to list alerts for dependent applications if relevant.

- **Dashboards**: Links to all dashboards, including dependent applications and systems.

- **Code repository**: A link to the Git repository the code is in.

Alerting

I often list my alerts in a table, so I can quickly identify by priority level what alerts have been created. It's also possible on some systems to link to search results for your alerts, which has the advantage of always being updated. You may also consider backing up your alerts, often available in tools such as JSON, or simply copying queries or making a screenshot.

Summary

Runbooks should allow most engineers to quickly learn about an application, including where to go when an application is having issues. We must remember that dependent applications and vendors can also be the root of an issue, so ensuring we understand those as well can be amazingly beneficial.

As we start generating alerting and dashboards to go along with our runbooks, one thing is clear – dashboards do not give us the ability to actively reach out to engage engineers, and alerts do not give a clear view of what is happening, only that something is wrong. Both are needed for the best engagement as a rockstar SRE.

Most importantly, this chapter talked a lot about priority levels and how to apply them. Many people are plagued with junk or false-positive emails and alerts – this sets the ground for alerts to be ignored, and worse, waking staff in the middle of the night without good reason causes poor job satisfaction and can increase staff churn.

As we continue our journey, we will explore outage management, from first engagement to closing the call. We'll share our years of emergency engagement routines, tips, tricks, and methods to make you a rockstar SRE.

14

Rapid Response – Outage Management Techniques

Responding to outages is anything but simple. We live in a world where technology outages can cause ripples in so many different aspects of our lives, from not being able to purchase gas to get to work, to traffic light outages impacting ambulance response times. When outages happen, the response has to be fast, organized, and have a well-defined process – or you'll spin wheels and eat up time trying to define simple things such as deciding whether to drive into the office or jump on a video chat.

There is so much more to outage response than just getting some engineers in a room to fix an issue – such as tracking timelines and determining the true customer impacts and metrics. Communication from outages needs to be concise and delivered at a steady cadence, with proper priority and impact information attached so executives can correctly understand the true severity of the outage. It also involves knowing when to send communications and knowing when not to communicate – when you shouldn't speak to both customers and leadership when small issues occur.

By the end of this chapter, you'll have a much deeper knowledge of the little things that can make responding faster, decrease the **Mean Time to Resolution (MTTR)**, and even help you walk through an outage, calling out both what was done right and what could be improved in our in-practice section. As a rockstar SRE, even the response to outages is an area to practice *constant improvement*.

Specifically, this chapter will discuss the following topics:

- Where to meet – an effective strategy for communicating good information
- Leveraging the people involved in the response
- The opportunity to respond at the right time
- Messaging customers and leadership
- In practice – applying what you've learned

We will start out discussing where and how to bring a team together and what impact it has on both the discussion and future historic information about the outage, which will be useful for generating the **Root Cause Analysis (RCA)**.

Where to meet – an effective strategy for communicating good information

A decade ago, outages were handled in a room of people, each poised with laptops and a cup of coffee, tackling the technical issue head-on. At night, this meant driving into the office and the lucky few found themselves on the conference call line active in the room.

In today's world of enterprise VPN and video conferencing, we often find ourselves joining in from our home office or dining room. Being able to be on a call in a minute instead of waiting for a team to arrive at the office isn't the only advantage. The impact on an employee's day or night is dramatically reduced as well. This type of near-instant response has become the norm in the industry.

Online collaboration

Often the online collaboration tools used at a company will pre-date you, and if you ever have the chance to weigh in on a new choice, platform capabilities nowadays are so similar, it may not matter for outage response communication.

The online collaboration capabilities of today allow for the rapid dissemination of not only messaging for conversations but we can also provide links to dashboards or cloud architecture console pages, and even copy and paste error messages and communications from a customer. This unwavering ability to bring together so much information can be an advantage, but can also overwhelm the team with unneeded information and distract from progress.

We must be mindful of the information we send out during an outage. Given the complexity of systems today, the scope should be kept as close to the issue as possible during the outage. I have even messaged others asking them to delete posts that were not relevant during outages.

Chat versus video

The big dilemma in today's world is the choice between chat and video communication, but I think the right choice is a balance of both. While voice communication allows a near-instantaneous response and the ability to share screens, chat history can make RCA creation much easier and retention policies allow us to keep those chats for months to look back at should issues re-occur. Chat also allows for a look back to see what may have been said about an issue earlier on – great for playing catch up for those joining late.

Most importantly, you'll want a strategy defined and agreed on before an outage occurs. This can include creating a new chat channel or video conference link per issue, the usage of chatbots to remind people of message cadence, or even instructions on the best way to copy and paste information into those channels.

In-person collaboration

There are, of course, those rare occurrences where online collaboration simply isn't robust enough, or remote access isn't possible. I have also seen teams be called to a location for longer-running outages or in response to political pressure from executives.

Regardless of the reason, when collaborating onsite, it's essential to back this onsite collaboration up with online chat. This allows the onsite team to quickly share information such as URLs or screenshots and aids in RCA information gathering later, which is often even more important because onsite outage remediation often has a higher impact radius or executive oversight.

All that being said, the best tip for remediating outages onsite is to ensure there is plenty of bottled water and coffee available to the staff – I have often even sent one of my engineers for coffee and water on the way to the office. Keeping the crew onsite both hydrated and caffeinated – and for longer issues, fed – is extremely important for the highest efficiency of the team.

The historical data found in outage responses

When we look back at outages, the data available in chat history and, if available, video recordings can give us a leg up on RCA data gathering. And by walking through the amount of data available post-outage, we are reminded of what we should document during an outage.

Timelines

Timelines are one of the most important parts of the RCA document. The basics of the timeline include the start and resolution time of the outage. But beyond this, we likely also want to know the following:

- Time to notify – the time from the start of the outage to the time the first alert or notification was made

- Time to acknowledge – the time from the start of the outage to the time the first alert was acknowledged

- Time to assemble – the amount of time it took for engineering assets to be available to start triage

- Time to remediation – the total time required to remediate the issue

Having this information available in a chat session will help you immensely when filling out an RCA.

Impact

Impact is a tricky thing. In my career, it has been difficult to judge impact based on what is happening, especially in a world with redundancy and queues. It is also greatly impacted by the customer, for example, if you know an event will sell out in minutes, does it matter if 10% of the customers can't complete their transaction when 100% of the tickets are going to be sold anyway?

In general, though, we look at impact through two lenses, **customer impact** and **business impact**. Some companies will place one above the other, but to me, both are good reasons to remediate an issue – and often customer impact has business impact ramifications.

Customer impact

Businesses are built on customers and providing them with a poor experience often drives customers away. It can also impact a brand and the customer's ability to trust the business. When an issue happens that causes an issue or poor experience for a customer, it is likely to cause churn – the loss of the customer – and will cost time and expense to acquire a new customer to replace them.

But here's the thing: customer impact isn't always what we think it is. If a customer can't log on to a website, it doesn't always have a negative impact – the impact comes when a customer obtains the services they desire regularly.

And it's not always about the customers themselves. Some services or offerings from a company are in such demand that poor experience will be tolerated in order to obtain an item or service. In these cases, we have seen in the industry that even large companies see no customer or revenue impact due to high error rates or high levels of poor customer experience.

We also must understand the simple question *who is the customer?* In terms of social media, it often isn't the end user posting on the page – it's the advertisers. So, if your advertisements are being seen and generating ad revenue, what impact does providing a poor user experience have?

In the end, taking care of a customer is important, but you must judge not only the customer but also product demand and repetition to truly understand customer impact.

Business impact

Business impact is typically a function of customer impact, as customers drive business. But there's not always such a direct connection and it is highly dependent on the business model. For example, a company that pays advertisers such as Google to bring people to their web page may have an immediate impact if clicking on the advertised link sends the user to a 500-error page. This is clearly a business impact – all that ad cost becomes a waste.

Some losses are less easy to spot, for example, not having attribution or link tracking enabled on certain pages can cause failure in seeing the complete picture of a customer journey, good or bad.

Of course, the biggest impact is often thought of as the lack of sales. But as you can see, it's not always about the sale – other items can clearly have an impact in unusual ways.

As we continue our journey into rapid response, we've discussed a lot about outages, so now we will focus on the people responding to the outage, as their involvement can be both good and bad.

Participants

It is important to note which people were involved in the remediation, either online, or made available to you in person. When it comes time for the RCA, knowing who responded is good for both being able to communicate about the follow-up after the outage and thanking the staff.

Follow-up work

During the outage, we should be aware of the future work needed. Often, we discuss *we should* items during the outage, and I encourage you to add them to the chat, as a record of items to remember. This simple step allows you to concentrate on the current issue at hand while releasing your mind from remembering what might come in the following days.

We have shown that the strategy for where and how you meet is vital to rapid remediation. Now, we'll journey into the people side of remediation to define types of typical work and how to position and task different personality types to bring about a calm and remediable recovery.

Leveraging the people involved in the response

People are what make the response, and leveraging every last person in the room can help bring down the resolution time of an outage. True, not everyone has the same set of talents, but all can be useful. When working in a team, ensuring not only that everyone feels valued but also feels they are making a contribution can keep people occupied and focused, which is better than sitting in a video conference making what-if proposals or complaining about the reliability of the technologies used.

Tasks

When we match tasks with users, besides getting the biggest bang for our buck in terms of remediation time, we must be very aware of the amount of support an individual may need, which can distract senior assets from their own assignments. I encourage you to start working with your junior engineers before an outage and make them very aware that during an outage, you – or other more senior engineers – may not be completely available to them. This is a great time for more junior engineers to learn, but they should be included.

Technical troubleshooting

We covered troubleshooting a few chapters ago and I encourage you to go back and walk through those steps regularly until you are fully aware of them. In addition, think about who would be the best person to be tasked with each of the methods during an outage.

Divide and conquer

This basic troubleshooting technique, often ignored, requires a diagram of the system plus the ability to break the system into parts. Starting halfway through the system, we test for valid data or activity, then continue to break the system in half again and again until we pinpoint the issue.

Previous catastrophe review

What happened in the past isn't always the best way to tell what's happening currently, especially given the continuous flux that changes systems, changing and morphing features both internal to our application and external to the third-party applications we interface to.

Looking at what happened in the past can be insightful, but systems tend to evolve, especially when the issue is specific to software. Absolutely take this approach, but be mindful, it may no longer apply.

Gut feelings

I know what it is, is a common phrase when some people dive into an issue – and truthfully, I say it myself. When we know our systems well enough, having a gut feeling will happen quickly but may not always be right. The one thing I'll point out is that engineers should go for the metrics and logs to prove their gut feeling – ensuring, during an outage, they prove their theory is important. Remember, poking around in the dark, while it may be fruitful, isn't an area where a lot of time should be spent.

Logs and code

During an outage, assigning someone to review the logs and possibly even recent code changes is always worth diving into. Log statements and links to specific lines of code copied and pasted into group chats can be invaluable and should be encouraged.

Communication

Assigning a single person to communicate with the rest of the company is imperative during the outage. You want someone who is trusted inside the company and understands the business and customer impact, as well as someone with strong communication skills. I often even redirect inquiries from departments to this person during an outage. As you'll find, the same message is repeated often.

Documentation

While going through an outage, if the capacity exists, having someone continuously gather the documentation can save time when writing the postmortem. It's not entirely essential to work through the outage but does have value if you can spare the capacity.

Busywork

I'm sure some are wondering how busy work made it on the list. It's because – and this may be the only book out there that says it – some people are valued employees, and we want them to feel valued, but during an outage, they add nothing and can even negatively impact remediation times. As the term suggests, this is work that has the value of removing distractions from the outage team.

Checking with the vendor

When working with third-party vendors, when they have an outage, it's not uncommon for communication to take a minimum of 20-30 minutes after an outage starts for the impact to be messaged out to their customers – if it's even communicated at all. Tasking someone with reaching out to vendors or even to do vendor-specific testing just makes good common sense – especially if we suspect the vendor is having issues.

Gathering customer impact lists

This task can be vital to outage recovery – especially if work needs to be done to resolve issues with data, however, it's also not always required. It is, however, very common – almost required – that we understand the number of customers being impacted.

Dashboard management

Assigning someone specifically to dashboards can have its advantage, but I often find many engineers update dashboards and alerts during an outage as part of troubleshooting. This is because we may want very specific insight into the outage, which allows us to make improvements.

It is also common that specific information may be needed about an outage, either while still happening or after, and building out a quick dashboard may be the fastest way to provide this insight. This creation of information can easily be assigned to a member of the remediation team.

Watch the number

Much like dashboard management, having a person who periodically checks the state of the impacted and non-impacted systems alike to ensure no other impact is happening is valuable. If you have the spare assets on the remediation call, ensure someone is taking a look into the operability of both the system that's down and other systems.

Participants and personalities

When a group of people comes together on a response, you'll find most are dedicated to the resolution of the issue. However, the impact they have on an incident can be widely varied – from being extremely helpful to slowing down, and even inhibiting, the resolution of issues. We'll walk through these:

- **The quiet**: Technical or not, those engineers who are timid, either by nature or fear of rejection, can be tasked on the side via chat and given both purpose and a channel to report back, though, this is most successful if you have previously established trust with the individual. This should not be confused with the absent type below.

- **The absent**: These are those who are on a call or on a chat and are otherwise occupied or not paying attention to the conversation. This person may be taking care of something personal, such as a child, or be completely disconnected. Defining the reason for the absence is imperative. If this person has family or personal obligations and can be released from the call, you should release them.

- **The talkers**: You know those people who just want to be heard, or worse, want the energy of the argument. I have always had luck in getting these people to make calls out to vendors and partners when they are involved, or tasking them with work that requires high concentration.

- **The incapable**: Those without the right technical or troubleshooting skill set often can participate without impact, though, for critical work, these people should be avoided, or if that's not possible, a review should be done of their work, or their work should be done via a screen share.

- **The in charge**: In a position of authority or not, you'll typically see one or two people taking charge in a call. Well-respected and knowledgeable members of staff with the capability should run calls. Our job then becomes one of a support role, not only helping when possible but also ensuring our own staff are not interfering.

- **The crazy ideas person**: Every call has its fair share of crazy ideas thrown out there, and the longer the call, the crazier things get. For this type of person, keeping a log of findings, returning back to them, and even updating them on why the crazy idea won't work, proves very effective.

- **The boss**: I've had C-levels on my calls refuse to take control of the call or make a decision of direction when two senior engineers are stuck on their own idea of a solution. I have in the past called this out on the call – well, the VP is on the call and has coding experience, let's see what they think. At the end of the day, I've found even VPs and CTOs sometimes just need a push.

Most personalities are mixtures of the above, and working with each type, especially in the high tension of an outage, can sometimes be difficult. You should think about, and even make notes about, different types of people when you are on calls. This will help in identifying them in future calls as well.

Overall, the job is to have everyone work together in a common direction. Don't be afraid to step in on a professional level, even if they are above you. I'm happy to shoot a note off to my CTO saying an engineer may not be responding as fast as they would like because they have three small children they have to watch too. Be alert, and most importantly, defuse conflicts and keep the work and the conversation moving forward!

Break strategy and stress management

The longer the call, the more stress it often creates – not only for the people on the call but for executives as well. Especially when the impact is directly tied to revenue dollars as opposed to issues with, say, existing customers not being able to see past orders, stress can decrease team effectiveness, which increases downtime and can cause confrontations on calls.

When you work on high-stress calls, remember to stay positive and help others do the same. Beyond this, these simple ideas will help with reducing stress.

Status updates

For those not on the call, or unable to follow along with the technical discussion, providing regular status updates has a great calming effect, as it provides not only a feeling that someone is working on the issue, but also provides information on the impact and that the resolution is progressing.

The worst place to leave your leadership is wondering what's happening.

Breaks

During a long outage, you must be mindful of the basic needs of the staff in attendance. Beyond just the basics of coffee or caffeine during the remediation, we must be mindful of much more. Using chat or messaging, checking up on your staff is imperative to ensure they don't need to step off the call. Knowing that they are far more likely to do so if leadership embraces the break, suggest to the senior leadership on the call that they announce a break.

Embracing a break while in full remediation mode is one thing, but I've always had great luck in discussing the need for breaks before my team even gets into a call. Making sure our crew understands that it's okay to take the time they need, even if it's to drop their child off at school, is important – after all, without the support of family, emergencies are even more stressful.

Task-based breaks

As a rockstar SRE, I've also used task-based breaks. It's easier to send a committed stress addict to the local bakery for coffee and a snack for the team than ask them to disengage for a personal break. And we all know those people who simply won't leave their team – for those people, we must be ready to task them in ways that show while they still remain highly valued, they can also take a few minutes to regroup.

Reducing overall stress

The notion of reducing stress overall is an important task of what we do in remediation and is especially important for high-impact or long-running outages. And this goes well beyond being nice and staying positive. Watching how others interact and being ready to be funny – or bringing in someone who is funny, can lighten the load. To reduce the stress of your leadership, ensure leadership are not only informed but feel they have an understanding of the issue(s) – offer to have someone spend time with them to help them understand the technical side.

Stress can elongate an outage and cause future political issues inside your organization. This is one time where ideas should be split into different paths of thought and troubleshooting, when we need to document what we've tried and identify, based on data, what causes we have ruled out.

The people involved in an outage response are very important, but knowing how to work with all types of people is also extremely important. As we continue on this journey of rapid outage response, we'll explore the opportunities we can find during that time.

The opportunity to respond at the right time

If it feels like we're talking about the people of the response and not how to make responses better, you'd be right. Understanding technology is about a third of response management – beyond just the knowledge and managing outages, always be mindful of the opportunities that may arise. Now, I'm not saying the outage should be extended so you can ensure better training or build out better runbooks, but when the company's talent is on a call and can contribute more to the technical conversation, you have the opportunity to do so much more.

Training

Training is always important. For those not aware of what the system is or its properties, watching higher-level engineers flow through page after page of configurations, code, and scripts will always teach a junior engineer something. It's often said that if you want to be smarter, hang around people smarter than you – and this is exactly that time.

To aid in training, I would try to ensure the engineer who is actively troubleshooting or remediating is sharing their screen and walking everyone through what they are doing.

You can also use time spent waiting when scripts are running or batches are completing, pulling other engineers onto the call to answer questions from other people on the call. Always embrace outages as teaching opportunities.

Dashboard and alert building

Our dashboards and alerts are never good enough. It seems like a constantly evolving and growing area of observability. In times when you're having outages, when you have to look up specific data, you have an opportunity to build a graph or add to a dashboard, or even to just share the query or link.

These simple steps, even just copying the query itself, can save time later on when you upgrade dashboards and alerts. Hang on to those things – post URLs in chats or send yourself an email. Later on, they'll make great references for building and refreshing observability.

Runbook and contact list revisions

Another great area of expansion during an outage is to increase the information inside runbooks. This information is vital to assist engineers who are not versed in products to learn and have easy access to information that may give them insight and wisdom.

Another great opportunity during an outage is to update contact lists, not only for internal employees but external partners who may participate in resolving the outage or may send emails during the outage – save this contact information, if only for availability to use it during an outage.

Team building

There is often nothing better than a full, all hands on-deck, outage to give people time to be together. Even in light of the stress of the outage, this time will build teams. From the comedic one-liner to discussing family and hobbies, this time together can be leveraged to build relationships.

Don't be afraid, during downtime, to have some small talk, and allow all who want to join in the conversation – and remember to ask the timid and shy what they think. Just be mindful that when the timeout is over, quickly pivot back to the outage.

Executive messaging bugs in the ear

There are many powerful ways to bring into the conversation areas of improvement that may be overheard by leadership. During an outage, making callouts for leadership to hear and understand must be done with a light touch. In fact, ensure the subjects brought up do not impact the call by casting doubt on the technical direction and reasoning unless you have a coalition of engineers ready to stand up and say the same.

In short, while you have the ear of an executive, it is often best to be mindful of what is said, especially given the complexities of outages. I recommend not making any callouts unless asked and then following up with your manager either later or by private message.

Opportunities to call out during the RCA

Finally, during the outage, if the chat is recorded, that is a great place to add statements to be reviewed by the RCA. In fact, I'll often type *"RCA NOTE ???"* before the idea and type in ideas for the RCA. Note the use of quotation marks – this is to ensure you give the impression this is not set in stone, but rather, a question or idea to be worked through later, ensuring the least stressful way to make these types of callouts in public.

Understanding both what else can be done during an outage and how it can grow both the team and provide insight into long-term stability is vital to continuous improvement – a true rockstar SRE calling. However, no outage would be complete without a discussion about communication, so we'll dig into that next.

Messaging customers and leadership

I will tell you from experience: I've definitely learned my lesson about messaging! Especially about messages to leadership and people outside of the technical organization. We'll talk about not only what messaging to send but how often, and the wisdom that not everything needs to be carried beyond your own boss or to be publicly announced.

Customer versus leadership messaging

I am a firm believer that the technology department in general should not craft the message to customers. Why? Technologists are not known for being the best communicators and they certainly don't spend their day being concerned about the brand and customer relationships. The PR and marketing departments are experts in understanding what the issue is, analyzing the impact on the customer, and then ensuring the message, if any, is communicated properly to the customer.

This is the same when it comes to executives! We can typically say something to our boss that would not have the same reception from their leadership – and the same applies to them. Unless I'm told otherwise, or I'm put on the spot by an executive, I always ensure my manager knows the ins and outs.

Cadence

During an outage, a lack of messaging can leave those not in the depths of the remediation feeling uneasy, and can even cause the outage to escalate in priority if messaging isn't done regularly. Now, my general rule is every 15 minutes for the first hour, then switch to every 20 or 30 minutes – and I would announce any change in when updates can be expected. You'll find within your own organization what cadence works, but always give contact information so interested parties can reach out. There is nothing worse than an executive dialing everyone, causing escalations and undue stress.

Email groups

I personally like email groups – especially when they are self-service. But be careful about the information put out in such a large group. As a suggestion, moderate your email groups, only allowing certain people to both send and reply to a group. This will both reduce noise and prevent small discussions from becoming extremely public.

Status sites

I'm not a big fan of "all-popular" company-wide status sites – and while I understand how useful they are, especially when managed by hand, they can be slow to show statuses and even slower to be updated.

If you go down this road, I will also caution against some types of automation that can trigger status changes. Make very sure your alerting is well thought out before connecting it to something that publicly calls out your company as impacted.

Over-messaging

How can you over-message? This I learned from experience: not everything is worthy of communication beyond your immediate leadership. Small issues, such as a lower-priority (P4) event, the lowest-priority (P5) event, or issues quickly remediated in production, such as a database being offline for only a few minutes, should not be messaged about. Why? Over-messaging leads to an overall trust issue with technology – partly due to the increased volume, but also because most people who are not directly involved with the systems impacted can have difficulties understanding what is a priority and what isn't. In other words, for every message, they think the technology is broken and will have a high impact – when quite the opposite may be true.

Even though I like to receive some messaging from others, I will always drop my own manager a quick message, starting with urgency, then the issue – for example, *"Blip but back up, database server failed over to backup, impacted for about 2 minutes during the switch."*

Notes, notes, notes…

By now, you can see the response isn't only about fixing the issue. There are so many equations to managing the people in the room, communication, and even the silver lining of opportunity inherent in outage response. For practice, we will walk through a response to an outage with a high impact on our business.

In practice – applying what you've learned

In this practice exercise, we are going to walk through an outage of our infamous custom order pricing API for our fictional custom hat website. We will start with the system going into a fault, and continue through alerting, engineers joining the call, communications, and final resolution. Like in the movies, you will have a better understanding of what's going on in the situation than those involved in the storyline.

To start, let's review the system itself – a web API designed to price custom hat orders as the customer goes through their journey of creating a design for their hat. Once each element of the hat is created, a backend system automatically evaluates and generates costing information. This method of pricing is far superior in that the calculations are half done when you go to price the entire hat, giving faster pricing feedback to the consumer:

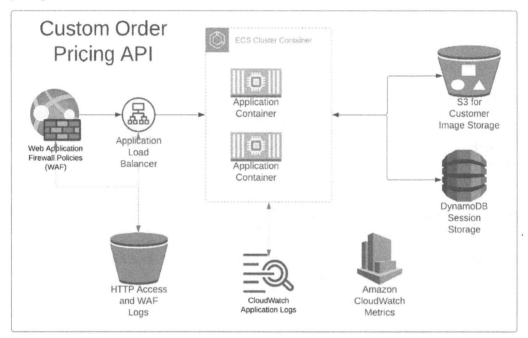

Figure 14.1 – Custom Order Pricing API Diagram

This system design is rather simple and should be found in our runbook – a proper **Web Application Firewall (WAF)** providing protection from illicit traffic, flowing into an **Application Load Balancer (ALB)** that distributes traffic to multiple application containers inside an **Elastic Container Service (ECS)** cluster that provides the application layer. Finally, we have **Amazon S3** and **DynamoDB** for data storage, and all of it is monitored in **CloudWatch Application Logs** and **Amazon CloudWatch Metrics**.

In this next diagram, we translate the architecture into its functional components or system view, identifying the backend APIs that exist inside the application containers and the relevant database tables and triggers associated:

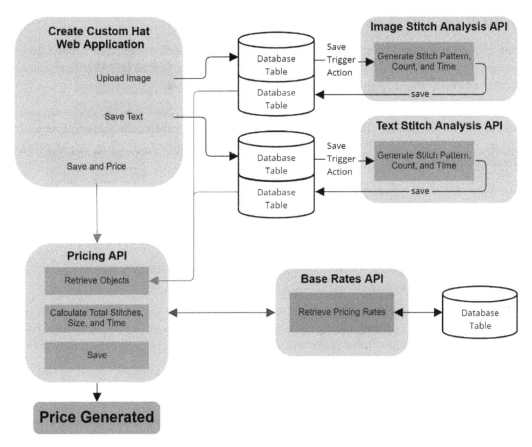

Figure 14.2 – Application flow diagram

When we look at how the application flows with respect to the rest of the system, we can see some very simple API calls happening while the customer is still creating their hat and the final pricing generation.

Outage and alarm

A recent mention of the company on a national news broadcast has brought in traffic loads of five times the normal load. When this happens, the DynamoDB table used to store the pricing starts to slow down, taking a minute, instead of a few seconds, to respond to queries. A custom CloudWatch metric tracks the amount of time it takes to calculate the final price, and a CloudWatch Alarm set to fire when the time exceeds 30 seconds. This alarm fires, sending a message over SNS to PagerDuty with a P2 priority.

Notification and response

When the notification is received by PagerDuty, the first engineer on call is paged. Ten minutes later, the engineer hasn't responded, and PagerDuty again pages the first engineer on call. After 15 minutes, both the first and second engineers are paged, along with the team lead.

After 17 minutes, the second-on-call engineer acknowledges the alert and joins the Zoom call and Slack channel for the outage – both already in progress. Two minutes later, the lead engineer also enters the Zoom call and Slack channel. Troubleshooting begins and an alert goes out to the company's technology-impact email list.

Thirty minutes into the outage, a call center manager, the web experience product owner, one junior developer, and two DevOps engineers join the call, and a few minutes are taken to brief everyone on the issues. After this, the lead engineer does a quick assessment of the people on the call and starts assigning tasks based on the type and capacity of the individuals. Here is that breakdown:

- **Second-on-call engineer:** A well-rounded, capable engineer is tasked with divide and conquering of the system in a systematic way.

- **Junior developer:** A junior developer is tasked with helping the second-on-call engineer by collecting screenshots and making notes of what is being reviewed, in both an assistive role to allow the engineer to move faster, and to provide a real-world training opportunity in the "divide and conquering method" of troubleshooting.

- **Product owner:** A web experience product owner is tasked with diving into Google Analytics to review web statistics and provide customer impact numbers. The task is designed with highly detailed requirements to prevent the product owner from injecting themselves into the technical discussion as they lack a truly technical skill set, yet insist on participating in the troubleshooting.

- **Call center manager:** A call center manager is tasked with collecting feedback from their staff and providing feedback to the executives since they are less dramatic than the product owner, which allows for better communication.

- **DevOps engineers:** One DevOps engineer tends to be rather quiet and reserved, so the leader messages them via chat and asks if they can review the CloudWatch metrics and logs and get back to them. The other DevOps engineer likes to call out every possible issue imaginable, jumping from idea to idea, so the leader asks for them to start going through the last 10 causes of known outages to the system, to validate none of them are the cause. Ten minutes later, the lead engineer hears a baby crying in the background and messages this engineer politely, and privately, giving them an excuse to leave the call – they accept and leave.

As we can see, there is a wide range of responders on the call, from the least experienced to the most experienced, from those who can provide reasonable feedback, to non-technical individuals attempting to appear technical. For anyone who's responded to calls in the past, this mix should seem rather normal.

Troubleshooting

While dividing and conquering the system, it is found that the text stitch analysis data in the database often cannot be found during the final pricing phase, causing the final pricing phase to generate this data during the intermediary pricing phase.

Upon further investigation as to why the text stitch analysis data is not in the database table, we find the calls to save to that database table are failing sporadically. Given the added load to the system, we can theorize that the write capacity of the DynamoDB table is being exceeded. Once adjusted, the problem seems to have lessened, but pricing times still remain at 40 seconds.

In reviewing the architecture further, we continue breaking down the system until we find the errors on the frontend: due to the wait times for the web layer to service calls, API calls to the application containers are timing out.

Eighty-seven minutes into the outage, the number of containers in the ECS cluster has increased and we see the times in the web layer return to normal. Teams stay on the call for another 10 minutes monitoring the website and creating a quick follow-up list for the next business day:

- DynamoDB cache?

- Auto scaling ECS cluster

- Can we auto scale write and read capacity?

- Add monitoring for DynamoDB capacity issues?

- Monitor the web page for timeouts?

- What could we add to dashboards and alerts to decrease the resolution time?

As you can see, most of these are questions and areas to follow up on. In addition, the lead engineer makes a note to talk with the primary on-call engineer.

The conclusion

In reviewing the issue, we can see that allocating tasks kept some of the responders on a path of resolution while keeping others from causing delays, and everyone feels like they are contributing.

And the primary on-call engineer? The lead engineer found out they were at the doctor's with a sick child and had messaged the secondary on-call engineer, which was both proper and appropriate.

In the end, the resolution is complete and a list of follow-ups has been created – no outage is pleasant, but this is a success.

Summary

No outage is pleasant, but when you get a group of responders together, in a high-pressure, high-stress situation, tempers will be tested. I suggest working with both your product and technology to ensure you build a response that allows those with the best abilities to remediate issues, even if that's a prior agreement to have a more senior engineer take the reins on demand for issues.

When we work with others in these situations, we must try to remain calm, focused, and objective. Assigning tasks can be both powerful in resolution and can offer a much-needed outside focus for those who are less likely to provide direct value in the technical response.

A final word on communication: having over-communicated and under-communicated at different times in my SRE career, finding the middle ground can be difficult. Listen to the buzz of the company and discuss this communication cadence and thresholds with everyone from product to engineering.

We'll dive into how we document and build future actions for outages in the next chapter. Not only is it important to describe the incident but to build blameless, fact-driven descriptions. We'll also work through how to identify future actions to help prevent future outages.

15
Postmortem Candor – Long-Term Resolution

The idea of the postmortem, also referred to as a **Root Cause Analysis (RCA)**, is to answer at least three key questions – what went wrong, how it was resolved, and what can be done to prevent it in the future. Often, this includes a timeline of events, information from systems and, specifically, their service level indicator, and how the service level objective may have been broken.

Understanding what went wrong is only half the battle; defining it in a way that allows neutrality and states facts only is extremely important. Events and data rather than emotions should drive our discussions. When we call out causality, it should be entirely fact-based.

As we journey through this chapter, we'll identify more than just the postmortem contents. We'll discuss how blame and opinion can cause a document to lose its effectiveness and make future proofing against the same issue more difficult. We'll also discuss how, sometimes, the cost to improve a system to prevent future events may not be in the best interest of a business.

In this chapter, we are going to cover the following main topics:

- The contents of the postmortem in executive summary style
- Decisions are not to blame
- The cost of more reliability as a business decision
- Training and skill sets – they matter
- Creating future action plans
- In practice – an example of a postmortem

We'll finish off the chapter with an in-practice session that takes what happened during the in-practice outage walk-through in *Chapter 14, Rapid Response – Outage Management Techniques*, and produces a postmortem.

The content of the postmortem in executive summary style

We'll start this chapter by going through many of the common elements of a postmortem or RCA document – I say *many* because every organization has its own ideas of what goes into this document, how it is stored, and what it calls out. I urge you to both understand how your organization manages theirs and, if you think it's incomplete, ask about changing the document. The most powerful thing we can do to prepare for outages is to prepare what will happen after, before it happens.

Executive summary style

Before we talk about postmortems, we need to understand the simple truth that the style in which we write has a high impact on how the document is received and the action in response to the outage. This style includes a number of basic ideas.

Simplicity

Do not dive deeply into the technical details when not needed, and ensure the ideas formed from the information can be easily understood by technical and non-technical audiences alike. In my career, I have also passed my RCA to a non-technical team member, typically a business partner or product owner, to ensure they understand the ideas of the document.

Connectedness

One of the things to think about is, "*post hoc ergo propter hoc,*" which is Latin for "*after this, therefore because of this,*" which is the idea of connectedness or causality. Take, for example, tripping on a power cord and unplugging a server. First, was the server already off or impacted? Is this really the cause?

And is tripping on the power cord an acceptable root cause? Or is the installation, or the failure to follow the process and procedure of a data center, the root cause?

There is nothing wrong with tying some events together; however, we should always attempt to remove potential blame points – for example, a power cord was tripped over, unplugging the power from the server, due to the improper installation of the power cable. Note how the statement still says the cord was tripped over but focused on the root cause of improper cable installation.

Numbers, metrics, and logs

When we speak of what happened, we should avoid a *story* but rather identify the numeric and log data that is available – for example, "*At 3:30 p.m., the CPU metric for the server started seeing 100% load.*"

Speaking in terms of provable items, such as metrics and logs, can provide indisputable proof of what happened, without much room to refute the incident findings.

Blamelessness

In order to keep the door to critique closed, we make every attempt not to call out blame. This can be tricky because, if someone makes a mistake, our attention naturally goes their way. The truth here though is if someone caused the outage, the responsibility to assign blame, or not, sits with the manager of the individual, not you.

The exception to this is, of course, admitting to the blame as your own, or your team's. This tactic should be used with caution but is effective in removing blame from all others.

Most importantly, if we see shortfalls in staff who may actually share some responsibility in the decisions leading up to an outage, we should identify how those decisions may have contributed to the outage but not attach that to individuals or teams.

Next, we'll discuss the overview of the document and how these simple few sentences have a high impact on the rest of the postmortem.

Overview

The overview should be a few sentences that simply describe the issue and how it was resolved. This should be as non-technical as possible, while still allowing technical people to understand the cause, and often includes a sentence that defines the impact on business and customers.

This overview should be able to identify everything those not looking for highly technical details of the issue want to know about the outage. In fact, this overview should be considered a simple description of the entire event.

Impact

Impact on customers and business is extremely important, as it sets the groundwork for the business-side decision of the reliability discussion and what future work should be done.

And while business may have an idea of what the impact is, revenue impact may not always be as impactful as first thought. It's important to also find a way to verify impact numbers. Often, I have found this is best done with metric or log analysis, although other sources include web analytics and existing database queries.

Timeline

The timeline should call out very basic key times during the outage. Specifically, I always include the following times:

- The start of an issue
- The start of an impact (if different)

- When the alert notification was sent
- When the responder engaged
- When the issue was identified
- When the issue was resolved

There are, of course, other items to add, including perhaps time frames of different remediation steps, times at which other departments are added as responders, and when failed resolution attempts are started. This last item, calling out failed remediation, should be done with care, but on long calls, I prefer to highlight remediation attempts, as it shows the continuation of the process.

Detailed technical description

This is one of the areas of the postmortem where you want to bring technical details to bear. Including technical information is paramount, as it's one of the largest reasons individuals actually read a postmortem. In this area, I suggest including both log and metric information.

When postmortems are being sent externally, care should be taken not to expose business-sensitive, proprietary, or confidential information. This area can also be used to describe the issues in a highly technical matter.

Response

While not found on all postmortems, making note of those who joined the remediation call and those who made themselves available to you, even if they were not needed, is paramount in helping management identify those who assisted in a time of need. This is important not only as the basis of a simple dispatch of "thank you" emails but also serves as a strong call-out, identifying those missing from remediations.

Resolution

The resolution part of postmortems is key, and while it should be kept concise and easy for non-technical personnel to understand, important details should not be left out. It's also important to look at documenting any scripts or commands used to remediate, as they often both provide a technical understanding of the resolution and the research material needed to build long-term solutions.

When we discuss resolution, calling out specific individuals may be important, only if they are a silo of knowledge – that is, they are the one or one of very few who have the ability or knowledge to remediate. This too is extremely important in future actions, and this *siloing* of information and skill set could also easily be called out as a root cause.

Future actions

There is a bit of finesse that should be taken when documenting future actions – and after the overview, this area is the second most popular place to read during a postmortem. There are a few basic types of future action categories, although this is far from a complete list:

- **Software**: We break software down into a couple of categories because software can have a bug that impairs proper functionality, but the functional design can also be deficient. The big difference here is that a bug often only needs a few lines of code to be changed to be remediated; however, if the software simply isn't designed in a way that handles all possibilities or operational needs and thus fails, we classify that as functional:

  - **Functional issues**: Different than bugs, this is when the software itself is operating as designed, but the design is flawed in some way. The major cause of this is a breakdown in functional needs and is most common when interfacing with other systems.

  - **Bugs**: Bugs are when software is not operating as intended.

- **Third-party integrations (internal and external)**: When third-party systems fail to perform as expected, either due to a change or downtime.

- **Architectural updates**: Making updates to architecture, whether in **Infrastructure as Code (IAC)** or manually configured.

- **System maintenance**: This addresses system configuration and maintenance issues, from hard drives filling up to improper timing of rebooting systems after installing updates.

- **Staffing concerns**: This is a rather large catch-all category, including a lack of proper development, system administration, and DevOps staffing, either from a quantitative or qualitatively perspective. I often point out training issues separately.

- **Documentation**: One of the most important call-outs possible, knowledge silos or limited institutional knowledge in one area can lead, in some cases, to exponentially longer outages. Knowledge silos are one of the largest enemies of any good reliability strategy.

- **Training**: Either in terms of being able to effectively troubleshoot or understand a system or software, lack of training can cause longer outages and impact decisions and directions during remediation.

As stated, this is not a complete list, and you'll find, over time, the correct balance of information for your audience. Next, we'll dive into how past decisions impact the postmortem.

Decisions are not blame

It's easy to say that we should have bought a new car instead of keeping the poorly performing car we have, but there may be challenges in saving money for a new car and higher insurance and maintenance costs.

This simple analogy is often what I use to explain how decisions could have been made differently. However, for every decision, especially when it comes to redundant hardware or more software development to build highly reliable code, there is a trade-off with a business's budget and the bottom line. In fact, a business may need to choose between paying employees and buying redundant hardware.

Business is business

The best description of business I have ever heard stems from risk analysis. Business is choosing what amount of risk we want to accept to take a chance to increase the value of the business, whether in the form of brand or revenue. In short, we choose how much money to bet on new strategies and technology, hoping we will win.

In that light, we as engineers should be highly aware of the cost of what we do, especially when it comes to the cost of using cloud technology.

Resource and time constraints

When a business lays out a budget, which equates to acceptable risk, we can find ourselves a bit taken aback by the lack of funding, not only in terms of cloud or computer resources but also in terms of development time. We also see this constraint creep into other aspects of projects, including quality assurance.

Monitoring

Monitoring is, of course, the most vulnerable when it comes to budget decisions, especially given the costs of **Software as a Service** (**SaaS**) solutions and the fact that these costs are often proportional to the number of transactions.

Many teams are getting by this by having their SRE or monitoring teams be part of the set annual costs of a division, instead of the developer cost model, which is basically work billed per project. This way, reliability is less susceptible to the per-project budget constraint.

Looking at overall reliability as a business question instead of a technology question can shed some light on some outages, although not all outages are due to poor budgets. When we choose to limit technology resources, a discussion should be had describing the possible impact, ensuring the entire scope of the risk is defined.

The cost of more reliability as a business decision

I wish every outage came with a six-figure remediation budget to shore up our code, systems, and architecture – trust me, you'd be hard-pressed to see this once in your career as a rockstar SRE.

The truth is that reliability costs – and cost often comes down to being a business decision. With that in mind, we'll not only discuss the technical side of these reliability options but also, more importantly, the cost of both outage time and budget.

Active:Active

A very basic and highly effective strategy is to create two copies of an architecture and put them in different data centers or regions. You use some type of load balancing that distributes load between the two. The downside is that for zero-impact outage deployments, you still require retries in your applications. Also, you have to have enough capacity online, often called **spinning capacity**, to accept the entire load in both data centers and regions, which amounts to running twice the number of servers you normally need.

Often thought of as one of the most expensive reliability options, depending on the architecture, this may be a great option. For server- or container-based architecture not running on serverless architecture, we often see high costs due to having multiple servers spinning at once, but for serverless architecture, where startup times take seconds, we can find a trade-off of using automatically deployed and scaled serverless architecture at very little cost increase.

The disadvantage of serverless is on the provider side; if a major data center fails completely, the load on the other data center's serverless capacity may become overwhelming, meaning you won't be able to start new serverless instances. In fact, as some serverless applications start cycling to new serverless instances, your entire application layer can begin failing in the remaining data center.

While the most expensive, cloud assets that are always available to you, called **spinning capacity**, provide the greatest protection from this type of complete failure – but at a tremendous cost.

Manual failover

As opposed to active-active, manual failover requires the change of some component, often a load balancer or DNS entry, to switch the load from one set of systems to another. As this is less costly, especially if the secondary system is set to a low computer capacity that can be scaled up on demand, it is often used. It should be noted that this methodology is also often found in older designs and is not as popular as it once was.

Cost of time to identify

Outage detection often requires some amount of time to identify. For containers, this may be the 3 minutes a container repeatedly fails a health check 5 times in a row, or perhaps the alarm that only fires if 20 errors happen in a 15-minute period. When we want to fail our systems over to a backup system, either manually or automatically, this time must be considered in the overall design and should be considered as downtime that will happen when failing over systems.

The cost of time to move a load

After we choose to move a load, either to serverless or architecture that has to be scaled up, there is a cost of time in moving that load. The time to move traffic to failover systems varies, based on the type of architecture. Of course, spinning capacity has far less startup time, and therefore, transfers load faster.

One common way of transferring load to failover systems is by changing the destination of a domain name by altering the **Domain Name Server (DNS)** record. DNS records include a setting that identifies the amount of time record caching is allowed, called the **Time to Live (TTL)**. Transfer of the load to the failover system would require waiting until the DNS records expire and the updated records are loaded.

For serverless architecture, you must allow time for new architecture to spin up, which can take from seconds to a minute or two, depending on many factors, including the image or application size, type of architecture, and calls to layers, such as credential stores the application must make on startup.

For autoscaling servers, the time can be as low as a few minutes, though a more reasonable time frame is in the 5–15-minute time frame. This is due to often having to reload the entire server OS and software.

Load balancer-based changes are often nearly instantaneous, but the automated failover is often triggered by a health check, which must expire multiple times before taking the application offline. This can take minutes, leaving load balancers to send traffic to non-responsive servers for that long.

In short, there is no architectural solution that is entirely foolproof and will switch over without a single glitch; they all have time requirements in which failure will occur. *Therefore, retries should be added to each layer of applications.*

Hidden development costs

The other cost involved in outages is the cost of the development time to not only diagnose and discuss remediation of the issues but also actually do the work. Longer day-long outages can easily impact entire sprints of work, especially with follow-up RCA discussions included in that timing.

I have always made this expectation as part of discussions about priority, to ensure everyone is not only aware but also in agreement that in the case of extremely high-priority (P1) and high-priority (P2) issues, an immediate response is expected from developers – even if that sets other projects back.

Beyond development costs, we also often look into the training and skill sets of the individuals tasked with being on call for remediation – and it can be expensive. We'll discuss that cost and the impact on teams next.

Training and skill sets – they matter

When I onboard a new SRE or DevOps engineer, I allow at least a 6-month ramp-up for them to learn the company's architecture, policies, people, and business. Can that time be faster? Certainly, even with a skill set in every tool used, training will still be needed for custom build architectures, deployment pipelines, and understanding of authentication systems, just to name a few. It's expensive to hire an SRE or DevOps engineer when you think of the time invested in training; I estimate it costs at least $50,000 to onboard an engineer in the United States market.

With this being said, identifying gaps and training opportunities quickly can allow the gaps to be addressed efficiently to ensure the best possible remediations.

Identifying gaps

So, how do you identify a gap? This may be as simple as just walking the engineer through something, but ultimately, it often requires one-on-one time with a senior engineer or technical leadership. And it can take time to understand these gaps, given the vast number of tools and knowledge an SRE or DevOps engineer has to be proficient in.

Training and certification targets

Training to be certified can be a quick way to identify basic training and targets for individuals and teams. Holding multiple certifications myself, I can tell you that I learned much during that study process and class time. The downside here is the cost, including the engineer's personal time, the classes and training materials, and of course, the certifications themselves. There is often an impact on productivity when assigning training, which should be discussed with your team to ensure both you and they are aware of the limitations of time assigned to training.

It should be noted that when I ask my engineers to get certified, we ourselves should also lead by example and ensure we have certifications as well. Doing so will certainly increase the probability of the rest of your engineers getting their certifications.

Cost is always a concern in a business; after all, it's business. Now that we've talked about what you can assign as future actions, next we'll dive into how to build that into a plan and the important call-outs to make.

Creating future action plans

Constant improvement, the basic rule of a rockstar SRE, should be the force driving plans for future actions. We also should be openly discussing with teams the risk of continuing without making these remediations, including product, development, and even security that is required. Be prepared to hear, "*no, we can't do that*" – sometimes, that's okay, but like anything involving politics, you must pick your battles. Not everything can be a priority; otherwise, there would be no priorities.

Immediate follow-up

These are items to be immediately followed up, such as bug fixes, alerting updates, and monitoring enhancements. In fact, when I add immediate follow-up items, I often talk about the small level of effort when applicable to bring simplicity to the requests.

To get the best engagement on these items, you must sell the *why* or the reason for the need; without doing this, expect leadership to answer "no" many times. Instilling business value in our presentations is the most important part of immediate follow-up items.

Who to involve

Before I write an official postmortem, I prefer to review the findings with my manager and a few of the trusted engineers I know in the organization to ensure that I have the proper bearings for the review, including any call-outs or items to remove from the document. Most importantly, I walk through my action items to ensure I have their support and judge the resistance to items I may find, allowing me to work on reducing that resistance. I also release the postmortem to my manager before widely distributing it.

I want to officially include anyone with a business or technical stake or interest in the issue and invite them to review and comment on the postmortem. I don't typically restrict access to the postmortem unless asked by my leadership.

For the individual action items, I prefer to let those reviewing the postmortem assign the work. However, there are times when only specific people should or can work on certain aspects of remediation, and these should be called out individually.

Timelines and priority

Even though this may be an unpopular option to some, I like to add timelines and priorities to the action items. Ensuring there is follow-up for work a group says should be done is imperative in the continuous improvement of systems.

I will often ask the assigned team leadership for a time frame in which to check back with the team, and the check-ins are best done in the presence of, or with updates being made to, leadership. Without this follow-up, ensuring items are completed can be an issue.

Assigning ownership

Assigning ownership will fall to a team; however, you may want to ensure it is worked on by specific people. The most important issue here is ensuring you follow a team's intake process for new work, even if the work was previously assigned as part of the follow-up discussions with the postmortem.

The importance of assignment is extensive, as it gives a focus point for status requests and ensures the work does not become a "hot potato," tossed around from team to team.

Tracking the work

Finally, you'll want to track the work. I prefer monthly updates for action items, so we keep the remediation ideas fresh in the mind, and after timelines have expired, there is a real choice to be made on whether an issue should be escalated or refreshed with the team, or whether it dies a silent death.

The most important part of the future action plan is its existence, with timelines, assignees, and of course, the ability to make people responsible for the work to be completed.

In-practice – an example of a postmortem

In this practice, we are going to build a postmortem from the outage walk-through from the in-practice section of *Chapter 14, Rapid Response – Outage Management Techniques*. We'll start with the all-important executive statement describing the outage and then continue to the other parts of the postmortem to complete the picture, including technical descriptions and action items.

Writing the overview

These few sentences, written in the all-important executive summary style and providing the best insight into the outage, are perhaps the hardest part of any outage.

Root cause

The cause of this outage was the inability of the website to handle the increase in customers visiting the site.

Technical failure

Autoscaling limitations were not set high enough to scale out the system properly, due to a high load.

Action items

Ensuring we have the proper capacity to move forward and the ability to respond faster is imperative. In addition, we need to add insight to the architectural layers to ensure the proper capacities are increasable to handle load increases such as this.

Overview statement

Now, we combine these items together to make a singular thought, and remember, we do not want to call out blame. Let's see our first attempt:

> *"Highlighting our company's continued commitment to our community, Channel 9 featured us on their recent broadcast for our work with the local kids' club. This increased the number of visitors to our website, a good thing, which our engineering teams were not prepared or capable of handling. Losing sales for at least an hour, they were eventually able to resolve the issue. Engineering has assured us that they will fix their capacity issues so they do not cause us to lose money should this happen again."*

So, what is wrong with this overview statement?

- The statement is simply too long and we get lost in details that are not relevant

- Multiple times, the engineering department is blamed – *"which our engineering teams were not prepared or capable of handling"* and *"so they do not cause us to lose money should this happen again"*

- The resolution is limited to a single department, *"Engineering has assured us,"* when this should be broadly discussed due to the cost of additional capacity

Let's try to rewrite the statement in a more concise, less blame-filled way:

> *"Our company's volunteerism was highlighted on Channel 9, which drove an unprecedented volume of traffic to our site, exceeding our provisioned capacity and slowing our systems down until they were non-functional. Our cloud-based technology allowed us to rapidly expand our capacity to fill this need, marketing sent out coupons to those users who may have been impacted, and product and engineering are already engaged in capacity planning for our future."*

First, notice the intentionally positive feel of the paragraph, which was a good thing. We made good decisions about being a volunteer-driven company, choosing a cloud presence, and even taking actions not only limited to engineering but also marketing. And we're working together to plan out what's next.

Rounding out the postmortem

The rest of the postmortem has the same feeling but has a few things to call out. The timeline does not leave large gaps in time but, rather, identifies multiple items worked on during the outage. The Impact section includes a note that some of the revenue could possibly be recovered, which gives a feeling that action is already taking place. And finally, we identify action items, not solid must-dos but calls for discussions, both with the cloud vendor and product, making the future actions not just technology-centric.

Next, you will find our postmortem for the Custom Hat Company outage, which offers a great reference for building your own postmortem. Most importantly, note the simplicity of the document – the best postmortems are simple and easy to read:

Custom Hat Company postmortem

"Our company's volunteerism was highlighted on Channel 9, which drove an unprecedented volume of traffic to our site. Our cloud-based technology allowed us to rapidly expand our capacity to fill this need, marketing sent out coupons to those users who may have been impacted, and product and engineering are already engaged in capacity planning for our future."

Impact

In total, 13,874 users experienced errors or were unable to complete the custom hat build process or check out. We estimate the lost sales volume to be approximately $187,000, though some money was recuperated by an immediate marketing campaign targeting the impacted users.

Timeline

This outage started at 5:37 P.M. and ended at 8:23 P.M. Here is the detailed timeline:

- 5:37 P.M. – Channel 9 aired a segment on our volunteerism, driving customers to our website.
- 5:49 P.M. – The alarm for pricing taking too long fired.
- 6:06 P.M. – The secondary on-call engineer responded to the call, started a bridge, and started investigating.
- 6:19 P.M. – The call center manager joined the call in progress, which included two DevOps engineers, the secondary on-call engineer, the product owner, and a junior developer.
- 6:37 P.M. – Additional containers were added to the web application layer
- 6:46 P.M. – Caching was enabled on DynamoDB.
- 7:24 P.M. – Read and write capacity was adjusted for DynamoDB.
- 7:37 P.M. – Caching was disabled on DynamoDB.
- 7:41 P.M. – The system was operating properly.
- 8:23 P.M. – Marketing sent out targeted emails, offering a discount to those impacted.

Technical details and response

The read and write capacity for DynamoDB is a limitation put in place in our database tier. Because the change added to the additional cost, the capacity was limited to triple our highest historic load on the website.

The scaling capacity of our web application layer was limited to prevent constant scaling up and down the layer, due to sudden capacity increases.

Resolution

Adjusting the method in which our application layer scales out and increases both the capacity of the database and application tiers resolved the issue.

Future actions

To increase reliability and help prevent future outages, the following actions are recommended:

- Discuss cost-effective solutions for database tier scaling options that may allow higher scaling at lower operational costs
- Work with the DevOps team to understand and adjust the web application layer autoscaling
- Define new cost estimates and review possible cost savings opportunities with the engineering and product departments
- Enhance dashboards to provide a better view of the database tier and capacity

Summary

Postmortems or, as some call them, RCAs, are vital to the outage process. They aren't only about the technical details of the outage; in fact, the technical details are actually the easiest and often the area of least concern – probably due to the highly technical nature of a rockstar SRE. Ensuring we call out the proper information without finger-pointing is key in delivering a professional postmortem.

As you walk through doing postmortems, pay close attention to the future work callouts, and vet those ideas well, not only with your manager but also trusted allies inside the company. Two of the unspoken key points of discussion are to take in concerns and build a coalition, before walking into postmortem discussions.

Remember that a postmortem's future work should be tracked and revisited periodically. It's okay for items to be dropped or postponed far into the future, but a discussion with the team, who hold the business risk, should be part of that decision. And finally, tracking postmortem future actions ensures the work is still in progress and doesn't ultimately get forgotten about.

In our next chapter, we dive into what it takes to both hire and be hired as an SRE. Beyond just interview questions and the true reasons for the answers they are looking for, our final chapter discusses what to look for in a manager and employer to ensure the best fit for the future.

Part 5 -
Looking into Future Trends and Preparing for SRE Interviews

This part provides a breakdown of different job positions all called **site reliability engineer** (**SRE**) in the industry – and the typical job responsibilities and requirements of being an SRE. Most importantly, we'll devote an entire chapter to the interview from the perspective of a job seeker and a hiring manager. We'll also discuss what makes a good company and position and share some random thoughts and inspiring stories that have led us on our journey.

The following chapters will be covered in this section:

16

Chaos Injector – Advanced Systems Stability

Nothing can survive for long without evolving. **Site reliability engineers** (**SREs**) are in an endless cycle of learning and applying as they become proficient in making systems resilient to incidents, downtimes, and outages. When systems are stable enough, SREs understand how to increase system trustworthiness by injecting chaos into it and checking weak links.

Two of the most unique and notorious practices inside the site reliability engineering domain are the **wheel-of-misfortune** game and **chaos engineering**; they can bring any system's reliability to a new level. This chapter starts by explaining these two techniques and how they work towards increasing a system's availability, resiliency, and reliability. Then this chapter conveniently demonstrates how to apply such methods by using examples and exercises around the topic.

In this chapter, we're going to cover the following main topics:

- Comprehending the wheel-of-misfortune game
- Understanding chaos engineering for reliability
- In practice – employing the wheel-of-misfortune game
- In practice – injecting chaos into systems

Technical requirements

At the end of this chapter, you'll find two different labs based on a **Node.js** application. We want to ensure you leave this chapter knowing how to practically implement the wheel-of-misfortune game and chaos engineering tests.

You will need the following for the lab:

- A laptop with access to the internet
- Node.js available on your laptop
- An account on a cloud service provider (we recommend **Google Cloud Platform** (**GCP**); you can create a free tier account here `https://cloud.google.com/free`)
- The `kubectl` tool installed on your laptop
- The `gcloud` **command line interface** (**CLI**) tool installed on your laptop (if you are using GCP)

You can find all files about this chapter on GitHub at `https://github.com/PacktPublishing/Becoming-a-Rockstar-SRE/blob/main/Chapter16`.

Let's dig into the game that trains SREs to respond faster to problems.

Comprehending the wheel-of-misfortune game

There's nothing more potent than discovering new abilities through a fun game. If any practice evolves into a routine with repeated steps, people will soon stop doing it. That's also valid for how SREs pull knowledge from their systems. Keeping them aware of the system problems and, most importantly, how to resolve them is essential for any organization.

The wheel-of-misfortune process settles two things. First, it ensures that SREs translate major incidents into teachable lessons. Second, it instills all the benefits of a gamification approach for knowledge transfer among team members.

In the following diagram, you can see a high-level representation of the wheel-of-misfortune technique:

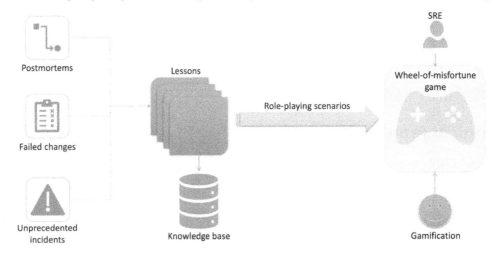

Figure 16.1 – The wheel-of-misfortune process overview

On the left side, we have a few possible data sources for the lessons we want to propagate throughout the team. SREs transform lessons into role-playing scenarios or playable simulation labs inside the game, as shown in the middle of the diagram. According to the right side of the diagram, a gamification strategy guides the incentives for completing any game stage.

We will divide the preceding process into four sections and provide detail on each:

- All ends are new beginnings
- Lessons to be learned
- Role-playing scenarios
- A little bit of gamification

We begin by looking at the inputs of the process.

All ends are new beginnings

After a production system outage situation, the sour taste in your mouth is not a good feeling for any SRE. We discussed the human element during an incident resolution in *Chapter 14, Rapid Response – Outage Management Techniques*, but what happens after we close a substantial incident ticket with a well-done postmortem analysis (a blameless one for sure)? You may say, get the incident's root causes fixed. Correct, and after that?

We want to extract facts from such failures and transform them into standard practices for site reliability engineering. Then later, we can mature the knowledge into individual SRE wisdom.

> **Important note**
>
> It's essential to differentiate what should turn into insight for SREs from what is just toil. For instance, if a change request fails because of a mistakenly issued command parameter, automating the change implementation (and eliminate that toil) is much more effective than teaching SREs not to make mistakes.

We first identify the data sources for the knowledge base of site reliability engineering; let's discuss a few possible origins.

Blameless postmortems

Incident postmortems generate many facts about the applications, infrastructure, and systems. We are interested in the facts that tell us how to diagnose a particular application problem based on its symptoms to the user. For instance, you need to make an HTTP call to the application API endpoint and check the transaction status when the latency is over a few seconds to determine whether there's a problem with an external dependency. The fact, in this case, is the correlation between the internal transaction status and the latency for the app user.

Other examples include detecting anomalies by monitoring multiple signals for a specific system and determining which workarounds we can implement to circumvent infrastructure component-level failures. All of the previous examples are runbooks containing facts, such as how monitoring signals normally behave or the best approach when a backing service goes down.

Incidents are not alone in this space. How about failed change requests? Let's review them as another possible source of facts.

Failed changes

Presumably, the most obvious place for finding facts is from procedures that didn't work well. After the root cause analysis is complete, what can the SREs learn from it? We can learn a novel or better method, technique, procedure, tool, or automation to perform operational work. Sometimes we search for shifts in how SREs handle the work to avoid user problems and impact as a collateral effect. Here are some exemplary facts from failed changes to illustrate this idea:

- Upgrading a **Kubernetes** worker node impacts all pods scheduled for it (Rod, co-author of this book, identifies with this one).

- A Kubernetes cluster update rollback or disaster recovery drill requires a full restore of the previous etcd database snapshot.

- Applications writing to a relational database may cause a lock if one is upgrading one of the related tables.

- There's no guarantee in which order an **Infrastructure as Code (IaC)** configuration creates resources without explicitly declared dependencies.

- Application manifest and container **software bill of materials (SBOM)** files contain all software asset dependencies. One of them may be compromised if not checked.

- Functions that read user data input or microservices that ingest data may be vulnerable to specially crafted data packets.

- Storage class and read-write policy significantly affect the persistent volume for a Kubernetes application.

Hopefully, you now comprehend how system or procedure failures turn into a source of facts (soon-to-be practices) for SREs. How about first-of-a-kind incidents? We talk about unprecedented incidents and all types of lessons to be learned in the next section.

Lessons to be learned

Often we call the knowledge extracted from a disaster a lesson learned, and little do we know it takes many other steps to learn that lesson. Undoubtedly only the people that experienced the failure event might have learned something. Soon enough, everybody will not even recall how they resolved that problem. To prevent losing precious expertise to oblivion, we need to transform found facts into yet-to-be-learned lessons through a simple process.

The first step is to rewrite each fact as a team directive. For instance, let's use the same examples from the previous section. Below we have possible team directives extracted from the previous listed failed changes:

- Do not upgrade a Kubernetes cluster worker node without draining it first if automatic upgrades are not an option

- Run an **etcd** database backup before any Kubernetes cluster upgrade and put it in a protected place

- Stop all applications using a relational database before upgrading any table inside it

- Include dependencies among IaC resources inside the same configuration file

- Scan application manifest and SBOM files for known critical security vulnerabilities (**Common Vulnerabilities and Exposures (CVE)**) before deploying code to production

- Drive a fuzzy test on functions or microservices that receive data inputs from external entities

- Properly document the storage requirements for any Kubernetes application

Second, we document those lessons in the knowledge base for the SREs by adding the situation and event details to help with the context of the lessons. Complementing them with information about those who were involved and having more information about the occurrence is also valuable.

As a third step, we determine the lesson variety. A lesson must purposefully lead to a change, and the essence of this change indicates what should be our course of action. The following are some of the possible outcomes of a lesson:

- **An additional task**: This change type is where we learned from past events that we need to add one or more functions to a procedure or runbook

- **Task modification**: Here, we must update one or more steps in a runbook

- **New practice**: This means SREs need to start a new approach in their engineering or operational work

- **Behavioral change**: This fourth change type means SREs must acquire a new ability, hone an existing one, or change how they do something

We check for toil elimination and automation opportunities regardless of the nature of the change. Typically, new practices and behavioral changes are good candidates for a wheel-of-misfortune game. Let's find out how we create such playable scenarios in the sequence.

Role-playing scenarios

Now that we have identified teachable lessons, we deploy them to our site reliability engineering team in an enjoyable way.

Role-playing, acting out, or pretending to be someone else to learn a new skill, is not new. We have employed this technique in dramatic plays, psychotherapy, and table or video games for a long time. It's no different for the wheel of misfortune practice. We conduct a role-playing scenario for the participants playing this learning activity where they pretend to be on-call SREs responsible for resolving the scripted problem. We create an equivalent environment with similar products, technologies, and applications as the production environment. Then, we configure it to have the same symptoms as the original event that led to this lesson. Then, we let the player figure out how to resolve this puzzle.

With Kubernetes, all you need to do is to recreate the production environment in a small-scale cluster and then set the role-playing scenario on it. If possible, you can also use a subset of production data to reproduce the situation with more fidelity.

You can spice things up by having a conference call during the game where participants play along with other people performing support roles such as the manager and customer. To make it fun, you can utilize more caricature roles. Rod will never forget when the fake executive actor in a role-playing game once said, "*it's the dog that wags the tail and not the other way around here.*"

Next, we understand how we make the wheel-of-misfortune game engaging and lasting.

A little bit of gamification

Gamification is the art of having fun doing work. We want to answer this question: why would SREs even bother to go through a wheel-of-misfortune game? What's in it for them?

Saying congratulations to SREs (the famous patting the back gesture) after they finish a role-playing scenario successfully is not enough. We use a gamification system to address the above questions.

The heart of any gamification system is the score. Engineers receive points for completing each step in the game and more points if they do that in the recommended way. Naturally, we design the wheel-of-misfortune game with multiple stages and possible ways to resolve it. We employ a ranking based on the score to generate healthy competition among SREs.

Later, the **human resources (HR)** department institutes a mechanism for trading points for products or benefits. Also, SREs earning a particular rank are entitled to a digital badge, shareable on social media. Now, who wouldn't be willing to go through a wheel-of-misfortune game?

As we write this book, gamification is reaching a new maturity level when companies release actual video games to teach technology. AWS, for instance, launched an interactive **role-playing game (RPG)** called **Cloud Quest** in 2022.

For reliability engineers and IT professionals in general, the importance of continuous learning and adopting new practices cannot be emphasized enough. Using a pragmatic approach with pinches of gamification philosophy, SREs can improve their skills through role-playing scenarios. Next, we talk about the ultimate testing for systems.

Understanding chaos engineering for reliability

Chaos engineering applies experiments to systems in production to find weaknesses and points of failure. Like fuzzy tests, a chaos experiment tries to break the infrastructure to understand how the system responds to the loss of a component. It may seem counterintuitive to introduce collapses to increase reliability. Still, if we consider it a specialized test, we want to pinpoint systemic flaws in a controlled manner instead of waiting to do that during a postmortem. It's called chaos engineering because we want to bring uncertainties from reality as variables of an experiment. Many IT parts, such as computing, network, and storage units, can fail, whether they are physical or virtualized. We simulate real-world chaos by deleting resources or shutting down components and checking how the system behaves. Naturally, we use a chaos system to manage and coordinate those experiments that inject chaos into a system.

To understand chaos engineering from a reliability perspective, we split this concept into three sections:

- Principles of chaos engineering
- Chaos system architecture
- Chaos experiments

We will start by looking at the principles of chaos engineering from its community.

Principles of chaos engineering

The chaos engineering community inside Google Groups published the *Principles of Chaos Engineering* manifesto. We recommend consulting it from time to time (`https://principlesofchaos.org`). The manifesto holds five advanced principles that should be part of any company's strategy for chaos engineering adoption:

- **Build a hypothesis around steady-state behavior**: This principle says to treat chaos exercises as scientific experiments where you must prove or disprove a theory
- **Vary real-world events**: Use meaningful and actual occurred failures as a base for new chaos injections
- **Run experiments in production**: Running experiments in production is the only way to obtain valuable insights into production infrastructure defects
- **Automate experiments to run continuously**: Chaos exercises cannot become toil and should run on new releases or infrastructure changes
- **Minimize blast radius**: We don't want to frustrate the system users too much, just enough to unveil faults

We would add the following two principles to the above list to accommodate the site reliability engineering approach:

- **Use the error budget to increase the frequency of experiments**: SREs cannot run chaos workflows as many times as they want. We can run a new experiment if there's a remaining error budget for the affected service level objectives.

- **Apply observability to explain the impact on the system's end user**: SREs don't monitor the infrastructure alone but adopt **four golden signals**, **application performance monitoring (APM)**, and **real-user monitoring (RUM)** to learn how an experiment affects the end user.

Let's look at chaos architecture to understand chaos engineering as a system next.

Chaos system architecture

Although breaking things is easy, doing that in a controlled way is not. We need a chaos system with execution and management functionalities to allow us to inject chaos and measure the outcome. We depict a typical chaos system in the following diagram:

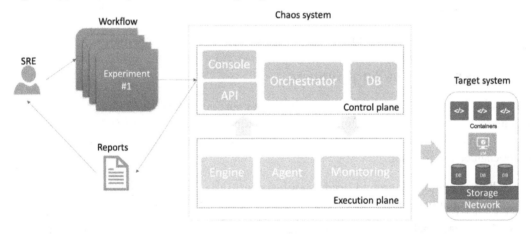

Figure 16.2 – Chaos system architecture

SREs write chaos experiments based on their experience from previous infrastructure failures. They merge experiments into a workflow using a console or programmatically via an API. Both entry points are part of the chaos system control plane, which is responsible for orchestrating the execution of workflows, managing reports, and storing results. On the execution plane, there's the engine or operator that injects chaos into the target system. A monitoring component is necessary to capture signals, traces, and logs from the system before, during, and after the experiment.

Next, let's find out what chaos experiments are.

Chaos experiments

We put chaos engineering with the wheel-of-misfortune game in the same chapter because they share continuous learning. Any new unforeseen infrastructure breakdowns may trigger the creation of a new chaos experiment. For instance, if a production database failover had a user-level impact, we should create a chaos experiment to see how the application survives a purposeful database shutdown. We cannot simply issue a full DB shutdown command; we want to ensure we're monitoring the system and disrupting only part of the services.

Although there's chaos in the name, the experiment is well designed, planned, and executed with fine control. Thus, we require a chaos platform and the following steps for a good-quality chaos exercise.

Characterize normal

Before proceeding with any exercise, we must comprehend normal behavior for the target system. We won't be able to detect anomalies or undesired behaviors if we don't know the normal parameters and the system's metrics baseline. Look at the golden signals and register the typical production latencies, error rates, traffic, and saturations. What should healthy transaction traces look like? And what are the regular quantities of certain logs?

SREs are ideal for writing experimentations because of their deep knowledge of system observability.

Create two distinct groups

Chaos engineering cannot create too much pain for the users; therefore, SREs must limit the blast radius of the experiment to a subset of users. Inside this division, we create two distinct groups: the control and experimental groups.

The control group contains the users that will not be part of the experiment. As in a new medicine trial, the control group encloses patients who take a placebo. For that, we target components not serving the system users in our control group. We can apply various load balancer routing rules to ensure no harm to our control group. Another possibility is splitting the incoming requests by user properties so we know which clusters serve a particular user type.

The experimental group has the users that will be subject to the experiment's effects. We inject chaos into the target infrastructure serving users belonging to this group.

Specify the chaos injection

Now that we know who the test subjects (part of the experimental group) are, it's time to determine what will fail. We can emulate a server crash, a Kubernetes pod deletion, a disk malfunction, or a cut network connection. Each type of failure will provoke diverse symptoms in a distributed system. Since we have contained the experiment to a group, we can simulate more than one failure. Usually, we mix a few exercises into a single workflow instead of running multiple experiments at different times – users tend to get upset if they feel impacted numerous times a month.

> **Important note**
> There's no sense in conducting chaos engineering on a system with a **single point of failure** (**SPOF**). If we disrupt any of those points, we will have a total outage, and the monitoring will capture no valuable data.

Check for observability augmentation

After we assemble experiments into a workflow, it's time to check for additional monitoring requirements. If we provoke a failure in a dependency, will the observability platform capture the data we need to analyze? Imagine spending your whole error budget on a chaos engineering test and not being able to seize the diagnostics.

Load and run the chaos experiment

Finally, we load the resulting workflow into the chaos engineering platform and schedule it to run. Since we are talking about production systems, we want to test a new chaos workflow in pre-production first. Of course, running any experiment on a staging environment will not give us the most helpful insights as we want to prevent weaknesses in production. Still, this testing of the chaos test will ensure the exercise doesn't break anything beyond the plan.

Compare groups results

Good chaos engineering tools provide detailed reports of the workflows and experiments. Also, since we have observability in place for production systems, we will have a broad spectrum of metrics, events, logs, and traces. We can then compare monitoring data from the experimental group to the control group users.

If the dissimilarities in system performance, resiliency, quality, availability, and reliability are minimal, then we have confidence that our production system passed the chaos engineering test. In a practical sense, meaning the system infrastructure is resilient, and the software running on it is fault tolerant. If not, SREs must work to improve the system design and reduce the identified defects.

Resolve fault points

The last step in the exercise is to eliminate systemic flaws. Although this effort doesn't belong to the chaos engineering discipline per se, it's natural to conclude experimentation with a solution.

As you may be able to predict, many possible system fault-fix pairs exist. Nevertheless, let's pinpoint the most common causes of failing a chaos engineering experiment.

Fault intolerance

Often the application or software is not sufficiently fault-tolerant, cascading errors up the chain to the user. The natural antidote for this flaw is the design patterns discussed in *Chapter 8, Reliable Architecture – Systems Strategy and Design*.

Architecture design

Other times, the cause of failing an experiment lies in the system design or architecture. Even though there's no SPOF in the architecture, it's possible to have components without a recommended number of instances, and those cases demand an architectural review.

Infrastructure configuration

The cause of this failure is the simplest one to sort out. Sometimes the platform or backing (backend) service configuration is insufficient to support consecutive component breakdowns. SREs adjust resources and infrastructure configurations as an action resulting from this chaos experiment.

Technical debt

As we transition from running workloads in on-premises traditional IT environments to cloud-enabled environments, companies are containerizing monolithic applications since complete refactoring would be cost and time prohibitive. Despite their positive application migration to cloud service providers, such workloads have intrinsic one or more SPOF and a lack of fault tolerance, making chaos experiments fail. Perhaps chaos engineering is not suited to those subjects.

Principles are the basis for any engineering discipline. Let's check chaos engineering's principles in order.

Next, we get into our labs to obtain some practical knowledge.

In practice – employing the wheel-of-misfortune game

As usual, we close the chapter with a hands-on experience with software tests in the testing simulation lab.

You will need the following pre-requisite knowledge to appreciate this lab:

- Familiarity with JavaScript and **Node.js**
- Basic understanding of containers and **Kubernetes**

We will divide this practical lab into the following three sections:

- Lab architecture
- Lab contents
- Lab instructions

Let's first begin with understanding the design for this lab.

Lab architecture

This lab uses a containerized Node.js application inside a **Google Kubernetes Engine** (**GKE**) cluster.

You must have the kubectl and gcloud command line tools installed on your laptop and the **kubeconfig** file (~/.kube/config) set for the GKE cluster under your GCP account (assuming you're going to use Google Cloud). We display the lab architecture in the following diagram:

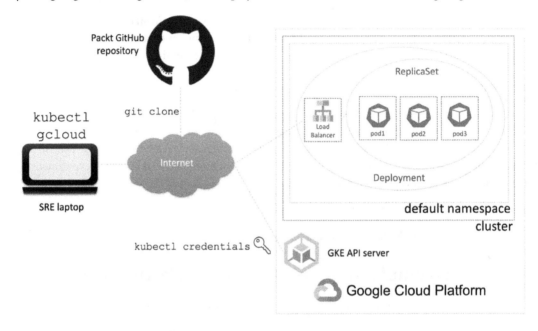

Figure 16.3 – Role-play simulation lab diagram

In the lab, we will deploy a simple Node.js app using the kubectl command and run scripts to configure port forwarding on your laptop. The application manifests have a load balancer service, a **NodePort** service, and a deployment with three replicas or pods.

We describe the contents of this lab next.

Lab contents

You can clone the GitHub repository by entering the following command in your terminal:

```
$ git clone git@github.com:PacktPublishing/Becoming-a-Rockstar-
SRE.git
```

Within this repository, under the Chapter16 folder, there are two sub-folders: chaos and roleplaying. We will use the roleplaying folder for this lab.

Inside the `roleplaying` subdirectory, there are three other sub-directories: `microservices`, `manifests`, and `solutions`. There's also a quick setup procedure called `roleplaying-simulation-lab.md` in the same place. This procedure details the installation of the CLI tools.

To access the application code, go to this directory:

```
$ cd Becoming-a-Rockstar-SRE/Chapter16/roleplaying/
microservices
```

To access the application manifests content, go to this another directory:

```
$ cd Becoming-a-Rockstar-SRE/Chapter16/roleplaying/manifests
```

Inside this directory, you will find the Kubernetes manifest files to deploy a simple Node.js app called `web-game`. Unfortunately, but intentionally, there are typos in the manifest files that will cause app malfunctions.

To check the resolution of this scenario, open the following directory:

```
$ cd Becoming-a-Rockstar-SRE/Chapter16/roleplaying/solutions
```

Next, we verify how to run this lab from your laptop.

Lab instructions

After you configure the GKE cluster with the information from your GCP project ID, it's time to run this lab.

We assume you already have installed and configured `kubectl`:

1. Go to the lab directory and then the manifests sub-directory. Use the following commands for that:

    ```
    $ cd Chapter16; cd manifests
    ```

2. Deploy the application to the GKE cluster using `kubectl`:

    ```
    $ kubectl apply -f *.yaml
    ```

3. Please wait for the `LoadBalancer` service to be assigned its external IP address. You can check the Kubernetes service with `kubectl`:

    ```
    $ kubectl get svc web-game-lb
    ```

You should see information displayed similar to the following:

```
NAME                 TYPE             CLUSTER-IP     EXTER-
NAL-IP      PORT(S)           AGE
web-game-lb    LoadBal-
ancer    10.108.15.145    34.176.33.152     60000:32391/
TCP    3m12s
```

4. You can run the following command to check whether the pods are running:

```
$ kubectl get pods
```

You should see the ImagePullBackOff fatal error message as follows:

```
NAME                         READY   STATUS          RES
TARTS    AGE
web-game-959899f46-f7s96    0/1     ImagePullBack-
Off    0          4m38s
web-game-959899f46-fmzb2    0/1     ImagePullBack-
Off    0          4m38s
web-game-959899f46-n579t    0/1     ImagePullBack-
Off    0          4m38s
```

Uh-oh, something is wrong. Can you figure out what's wrong with the web-game-deployment. yaml file? You can obtain more information by describing any of the pods:

```
$ kubectl describe pod <pod-id>
```

Change the web-game-deployment.yaml file to fix it, then delete the previous deployment object and create a new one:

```
$ vi web-game-deployment.yaml
$ kubectl delete deployment web-game-deployment
$ kubectl apply -f web-game-deployment.yaml
```

Pro-tip note

You can use the following kubectl command to change the image of a deployment object:

```
kubectl set image deployment/web-game web-game=$image:$version
```

---

5. Try to open the application from your laptop browser using the load balancer service IP address and port:

```
http://<load-balancer-ip>:60000/
```

You should see an empty screen while the browser waits for a response. If you wait sufficiently (with patience), it will return a connection timeout error like this one:

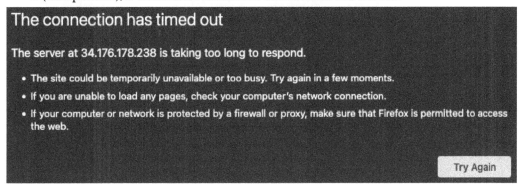

Figure 16.4 – Connection timeout error in the browser

Can you get to the application through the NodePort service? Establish port forwarding to link your laptop localhost to the NodePort service in the Kubernetes cluster:

```
$ kubectl port-forward svc/web-game-svc 8082:8082
```

You should now be able to open the tunnel in your local browser:

```
http://localhost:8082/
```

So, what's wrong with its manifest file if the NodePort service is working but not the load balancer one? Can you find the error?

6. Change the web-game-lb-service.yaml file to fix it, then delete the previous LoadBalancer object and create a new one:

```
$ vi web-game-lb-service.YAML
$ kubectl delete service web-game-lb-service
$ kubectl apply -f web-game-lb-service.yaml
```

> **Pro-tip note**
>
> You can use the following kubectl command to patch a service:
>
> ```
> kubectl patch service web-game-lb -p '{"spec":{"selector":{"app":
> "web-game"}}}'
> ```

You have completed this lab. Congratulations!

If you could not pinpoint the typos in the manifest files, don't worry, you will find the correct ones inside the `solutions` sub-directory. You can also run this lab in other *hyperscalers*[1], such as AWS and Azure.

Let's switch the context and check the chaos engineering lab next.

In practice – injecting chaos into systems

Since we talked about two distinct but correlated methods, we added a second simulation lab based on chaos engineering. This lab relies on the **LitmusChaos** framework, which has one of the best chaos testing systems for Kubernetes deployments. We deploy LitmusChaos to the Kubernetes cluster as **Custom Resource Definitions (CRDs)**, which is why we particularly like it.

You will need the following pre-requisite knowledge to appreciate this lab:

- Familiarity with JavaScript and Node.js
- Basic understanding of Kubernetes, **operators**, and CRDs

We divide this second practical lab into three sections as usual:

- Lab architecture
- Lab contents
- Lab instructions

Let's begin with understanding the design for this chaos simulation lab first.

Lab architecture

This chaos engineering lab deploys the latest stable LitmusChaos operator to a GKE cluster and a simple Node.js web application. It would be best to have the `kubectl` and `gcloud` command line tools already installed on your laptop and the kubeconfig file (`~/.kube/config`) set for the GKE cluster under your GCP account (assuming you're going to use Google Cloud). As usual, we illustrate this lab architecture in the following diagram:

Figure 16.5 – Chaos simulation lab diagram

We configure the LitmusChaos `chaos-operator-ce` component in the `litmus` namespace with three types of CRDs: ChaosEngine (`chaosengines.litmus.io`), ChaosExperiment (`chaosexperiments.litmus.io`), and ChaosResult (`chaosresults.litmus.io`). Also, we deploy an operator as a deployment object called `chaos-operation-ce`.

In the default namespace, we deploy a simple web app that we use as a target for the chaos experiment. We install a ChaosExperiment called `pod-delete` on the `default` namespace. We create a ServiceAccount to give it privileges for pod deletion.

> **Important note**
>
> We are not deploying the new Litmus ChaosCenter with an entire control plane in this lab. Instead, we are only using the `chaos-operator` component.

We describe the contents of this lab next.

Lab contents

You can clone the GitHub repository by entering the following command in your terminal:

```
$ git clone git@github.com:PacktPublishing/Becoming-a-Rockstar-
SRE.git
```

Within this repository, under the `Chapter16` folder, there are two sub-folders: `chaos` and `roleplaying`. We will use the `chaos` one for this lab.

Inside the `chaos` subdirectory are two other sub-directories: `node` and `litmus`. Also, there's a quick setup procedure called `chaos-simulation-lab.md` at the same level. This procedure details the installation of the CLI tools.

To access the application manifest files, go to this directory:

```
$ cd Becoming-a-Rockstar-SRE/Chapter16/chaos/node
```

Inside this directory, you will find the Kubernetes manifest files to deploy a simple Node.js app called `web-app`. There are three files inside this sub-folder to create a deployment, a NodePort type service, and a load balancer type service for the application.

To access the LitmusChaos operator manifests content, go to this other directory:

```
$ cd Becoming-a-Rockstar-SRE/Chapter16/chaos/litmus
```

Inside this second directory, you will see the Kubernetes manifest files to configure the LitmusChaos operator. The `web-chaos-sa.yaml` file creates the ServiceAccount used by the ChaosEngine to consult the information about the application and delete its pods. In comparison, the `web-pod-delete.html` file instantiates a new ChaosEngine that will run ChaosExperiment `pod-delete` on the target application.

Next, we demonstrate how to run this lab from your laptop.

Lab instructions

After you have configured the GKE cluster with the information from your GCP project ID, it's time to run this second lab.

We assume you already have installed and configured `kubectl`:

1. Go to the lab directory and then the manifests sub-directory. Use the following commands for that:

    ```
    $ cd Chapter16; cd chaos
    ```

2. Deploy the application to the GKE cluster using `kubectl`:

    ```
    $ kubectl apply -f node/
    ```

3. Please wait for the load balancer service to be assigned its external IP address. You can check the Kubernetes service with `kubectl`:

    ```
    $ kubectl get svc web-app-lb
    ```

You should see a piece of information similar to what is displayed here:

```
NAME        TYPE         CLUSTER-IP    EXTER-
NAL-IP  PORT(S)        AGE
web-app-lb  LoadBal-
ancer   10.36.11.169   34.176.89.96   60000:30543/TCP   58m
```

You can test the application here:

```
http://34.176.89.96:60000
```

4. You can run the following command to check whether the pods are running.:

    ```
    $ kubectl get pods
    ```

5. Install the LitmusChaos operator in a separate namespace, and issue the following command to accomplish that:

    ```
    $ kubectl apply -f https://litmuschaos.github.io/litmus/
    litmus-operator-v2.14.0.yaml
    ```

6. Check whether the manifest has created all CRDs, API resources, and the operator in the `litmus` namespace:

    ```
    $ kubectl get crds | grep chaos
    $ kubectl api-resources | grep chaos
    $ kubectl get pod -n litmus
    ```

    You should see three CRDs, three API resources, and one operator running as a pod.

7. Create a ServiceAccount to assign privileges to a ChaosEngine on the default namespace. Apply the following manifest:

    ```
    $ kubectl apply -f litmus/web-chaos-sa.yaml
    ```

8. Deploy the `pod-delete` ChaosExperiment from the **Chaos Hub**. Similar to **Docker Hub**, anyone can share a chaos experiment there:

    ```
    $ kubectl apply -f https://hub.litmuschaos.io/api/
    chaos/2.14.0?file=charts/generic/pod-delete/experiment.
    yaml
    ```

9. Create an **annotation** for the deployment application to mark it for chaos. A ChaosEngine will only inject chaos into an application by running a ChaosExperiment if it has this annotation:

    ```
    $ kubectl annotate deploy/web-app litmuschaos.io/
    chaos="true"
    ```

10. Create a ChaosEngine based on the `pod-delete` ChaosExperiment:

```
$ kubectl apply -f litmus/web-pod-delete.yaml
```

11. Verify whether the pods are being deleted by the experiment:

```
$ kubectl get pods
```

You should see something similar to this:

```
NAME                      READY  STATUS          RESTARTS  AGE
pod-delete-i0g5c1-bxgdt   1/1    Running         0         15s
web-app-6bd549dfb5-bjt6b  1/1    Running         0         15m
web-app-6bd549dfb5-c8frt  1/1    Terminating     0         15m
web-app-6bd549dfb5-j94kv  1/1    Running         0         15m
web-app-6bd549dfb5-
rrpf7 0/1    ContainerCreating 0         3s
web-chaos-runner          1/1    Running         0         16s
```

12. Finally, check the report on the ChaosResult resource:

```
$ kubectl describe chaosresult web-chaos-pod-delete
```

A good result looks like this:

```
...
Spec:
  Engine:      web-chaos
  Experiment:  pod-delete
Status:
  Experiment Status:
    Fail Step:                 N/A
    Phase:                     Completed
    Probe Success Percentage:  100
    Verdict:                   Pass
  History:
    Failed Runs:   3
    Passed Runs:   1
    Stopped Runs:  0
    Targets:
      Chaos Status:  targeted
      Kind:          deployment
```

```
        Name:              web-app
   . . .
```

You have completed this lab. Congratulations!

While the ChaosResult report is helpful in understanding whether the application has passed, or failed, the chaos engineering test, it's critical to have functional monitoring to capture changes during the experimentation. Many other experiments are available in the Chaos Hub, and you can also run this lab in other hyperscalers, such as **AWS** and **Azure**.

Next, we summarize what we have learned in this chapter.

Summary

SREs attain advanced stability levels by actively learning from system problems that are within their scope of responsibility and from purposely induced disruptions. SREs can gain wisdom through the wheel-of-misfortune practice that adopts a gamification approach. Chaos engineering is the ultimate testing method for mature systems where all other tests have not revealed deficiencies.

At this point, you should be able to conduct a wheel-of-misfortune game implementation in your organization and guide SREs on continually learning from the most noteworthy incidents. Capture the precious lessons learned from **blameless postmortems** and retroactively feed them into the knowledge base so the team can respond quicker in future similar scenarios. Remember to make all of those efforts a fun game.

By now, you understand why chaos engineering works and how you can apply it to systems to unveil hidden faults.

In the next chapter, we will coach you for an SRE job interview and discuss how to hire potential rockstar SREs.

Further reading

- We recommend the following article about gamification that Rod wrote some time ago, but it's still relevant: `https://www.linkedin.com/pulse/gamification-art-having-fun-doing-work-rod-anami/`

- To learn more about the AWS Cloud Quest RPG, please access the following page: `https://aws.amazon.com/training/digital/aws-cloud-quest/`

- If you're interested in learning more about digital badges, please visit Credly at this link: `https://info.credly.com/`

- You can read the full *Principles of Chaos Engineering* manifesto at this link: `https://principlesofchaos.org/`

- Learn more about the Litmus Chaos Engineering Framework here: `https://docs.litmuschaos.io/docs/introduction/what-is-litmus`

- You can check out the details of the `pod-delete` chaos experiment on this page: `https://hub.litmuschaos.io/generic/pod-delete`

## 17

Interview Advice – Hiring and Being Hired

Hiring in general is always troublesome. From a candidate's viewpoint, it's a struggle to bring your knowledge, skill set, and drive to get work done into the full view of the hiring team. On the hiring team side, we know the job of a rockstar SRE sometimes involves holding on with two hands as the world swirls around us in outages and team interactions – not always pretty in nature. Connecting from either side can be immediately great or immediately *blah*.

As we go through this chapter, we won't just talk about the questions you're likely to be asked and dive into the meaning of those questions, as not all are clear in intent. We will bring the ideals of what a rockstar SRE does into the fold of interviewing, discussion outages, and experience. We will also look to understand what the team needs and what is needed from you.

From a hiring manager's perspective, you'll enjoy a fresh crop of questions diving into the heart of more than just talent. We'll walk through finding passion and drive in a candidate. In a world where personal responsibility can sometimes be a challenge, we'll discuss how to drill down for more than just the technical details.

In the end, there is always an X factor, and we'll discuss how that edge can be advantageous or get you the quick rejection. I'll also share my experience and why sometimes, five minutes into an interview, I'm bored and start getting distracted – spoiler, it happened when I interviewed one of the best engineers I've ever hired!

In this chapter, we'll discuss the following key topics:

- What we're looking for in a candidate
- Common interview questions and answers
- What you should look for in a career
- Researching the company

- Are you over- or under-certified?
- Tips for landing a job with a great salary

What we're looking for in a candidate

SREs are some of the hardest to hire, as the job requires a rather large skill set, uncommon troubleshooting, and rapid learning capabilities, mixed with the ability to be responsible. You'd be surprised in the marketplace just how hard that combination is, especially when that skill set often includes development and cloud infrastructure.

Are you qualified?

This is a hard question to answer, and the entire reason for writing the chapter is to help you understand the types of experience you need in order to obtain a position as an SRE. Being qualified is both a mix of what a company is looking for and frankly, what they are willing to pay.

Entry-level SRE job

Let's start with this: SRE is not an entry-level position, though I have seen and even hired entry-level engineers; these are not the norm, and trust me, the entry-level pay is often half that of a seasoned SRE. So, if you are an entry-level engineer, where do you start? Simple, learn to code – and learn about the architecture it runs on. Find out who has the knowledge where you are and learn from them – be curious, be willing to grab lunch with senior engineers, ask questions, and above all, never stop learning.

Problem-solving

First on the list is problem-solving, as that's one of the major roles of the SRE. And it's more than just breaking down and fixing issues; along with this comes attention to detail and remembering what happens around the systems so that you can bridge gaps between what is happening and what should be happening. In fact, the entirety of *Chapter 5, Resolution Path – Master Troubleshooting*, is dedicated to understanding how to resolve issues.

The ability to accept feedback and direction

Building in isolation is never as complete as building something with multiple minds at work. The same is no different in site reliability engineering. Factor in that, at times, leadership will make choices that impact your work and sometimes, you're not given an explanation for why the work needs to be carried out in a certain way.

Part of this is being able to set aside your ego and technical preferences to bring into the mix the technologies and methodologies common to a company. It's important, in fact, extremely important, for all members of a team.

A broad knowledge base and skill set

Because issues can creep up in code, architecture, systems, and processes, it's imperative we know as much as possible when stepping up to bat. This means bringing people on board who have a wide knowledge base. Often found from years of experience doing many things, or working in small shops, finding someone who has done more than a single job or two, who has a passion for and went out to understand more than just code, that's the passion we're looking for in rockstar SREs.

Research and learning skill set

Having experienced many issues where Google has provided me with the answer, understanding how to quickly research and learn new things is critical in an outage. Being able to reach beyond our own knowledge and skill set makes us versatile and ensures we can be counted on.

I've interviewed a fair number of engineers who have antique skill sets, incomplete and years out of date; unfortunately, the skill set must also be relevant for the job. Not understanding cloud infrastructure or **infrastructure as code (IaC)** would be an instant no for my team as it's a core part of what we do. You'll find many who have a skill set gap and some who won't know the specific software or tools you may be using; this gap may be acceptable, but larger gaps, such as modern serverless architecture, may be a deal breaker.

The ability to say "No"

Here it is, the part of the job that requires us to be gatekeepers and say *No* to engineers. Being able to do this, and understanding its unpopular nature is important. Not all people have that in their wheelhouse.

Culture fit

Do you jive with your co-workers and workplace? Do you want to be part of a start-up with long hours and a large range of responsibilities filled with passion and drive? Or are you looking for a 9 to 5 job with a standard desk, or do you want to work from a bean bag with your laptop?

Can we laugh together? Can we listen to each other? Are we okay with the level of honesty you'll be getting and putting out there?

Culture goes well beyond the ping pong tables and free cappuccinos; it's how well you'll be able to fit in the workspace, with the team, and with your boss.

The X factor

If you think culture is a wildly undefined topic, the X factor is everything else. Often, it's the little push to yes or no that often can't be explained or is a preference due to the past experience of the hiring manager. I've been hired because my room is full of 3D printers and other such toys behind me in my tech cave, and I've not hired others because they smoked during an interview. These are all X factors.

But here are the two largest X factors in the industry.

Military service

Military service is a strange animal; most former military personnel are oddly slightly embarrassed when thanked for their service; for those of us who have served, it's hard to define what it is to those who haven't. Beyond this sense of belonging with veterans, there is an understanding of what leadership and hard work are – a sense that we know you made it through and can do the impossible when given the right support and tools.

University

This one is more about community and belonging than anything else, while some will say having a degree shows you can complete something. I can't say I completely understand it, having gone into the military instead of heading to university, but it's one of the most common X factors.

Passion

There is nothing that makes me smile more in an interview than a candidate who talks about a subject with passion! It could be the passion for building or remediating an issue, or even about a specific technology. I want to see it; I want to feel it, and I want it to drip from every word you say.

Experience

At the top of our list is experience. The SRE trade is rarely an entry-level trade, often requiring development and DevOps-type backgrounds and experience. Is this always true? Certainly not; I've hired many entry-level engineers into SRE and DevOps teams for assistance with development and work that is reparative and often boring – but with a great opportunity to learn.

Your experience should be broad; we want those who can not only discuss how to write an API but also understand the architecture of the layers in front of that API. While each chapter in this book provides a direct interface into our decades of experience, before you go to an interview, we urge you to review the following chapters. *Chapter 2, Fundamental Numbers – Reliability Statistics*, brings light to the **service level indicator** (**SLI**), **service level objective** (**SLO**), and **service level agreement** (**SLA**), which are core in the constant improvement methodology of an SRE. Here, we also define that it's not technology that defines the SLA but rather, the SLA is an agreement from both the product and business side with the technology side of the company.

Chapter 4, Essential Observability – Metrics, Events, Logs, and Traces (MELT), drives home the use of metrics, logs, and tracing in the analysis of the health of a system. Understanding how to validate what is and what is not healthy is key to outage remediation.

Chapter 5, Resolution Path – Master Troubleshooting, is an entire chapter about the processes employed to solve issues – and we want to hear about your process and your path more than just that you fixed this issue in an interview.

Chapter 8, Reliable Architecture – Systems Strategy and Design, walks us through the definition of reliability from an architectural standpoint. And most importantly, it helps you understand how the correct architecture can add resiliency to an application.

Personal responsibility

And here we are at the last item – and it's huge, especially in today's job market. We all know of people who cannot make it to work on time, are always missing meetings, or worse, are unable to take responsibility. The common term these days is **adulting**.

As a manager, I've run the gambit of issues with staff, from missing work because they slept in to simply not being able to stay on task or get work done, not doing training, or even arguing with their boss over something trivial.

I have known amazing engineers who simply could not function reliably and stay on task to get the job done. To me, the most important skill set you can have in this area is to take a problem to resolution within the confines of the tools and processes available in a company while keeping your manager in the loop on progress.

Next, we'll dive into common interview questions and answers. We'll also review what is being looked for in the answer.

Common interview questions and answers

Interviews are about discovery, and questions stand as the tool to bring to light the experience and skill set a candidate and potential employer have. We've separated the questions into technical, non-technical, and one final group of insightfully odd questions meant to dive in and retrieve specific answers while, on the surface, they seem odd and even useless.

Technical questions

Long gone are the days of simple questions and answers in the technical space. The technical skill set requires a larger and larger understanding of systems, not just components. For an SRE, this is even larger. These are some of our most popular question topics.

How HTTPS works

The transactions required to submit and retrieve data over an encrypted system are rather complex, requiring multiple calls back and forth between the end user and the target system or systems.

For this answer, I'm not so much interested in certificate revocation or specifics on the actual math behind the encryption, but an understanding of how private and public certificates interact with a certificate authority is key. Also, I'm looking for the multiple calls, what is and what is not public about the request, and if you throw in a bit about man-in-the-middle attacks, it's bonus points.

Infrastructure as code

Understanding how IaC works and having had experience in it is critical to environments that take full advantage of IaC. Typically, in the form of CloudFormation or Terraform, the end results and components tend to stay the same.

In this answer, I want you to first walk me through how they work, including state, and which components, such as security and permissions, are always required in building new architecture. In Terraform, I'm looking for explanations of state files and why they are good and bad. And bonus points for bringing up drift and how it can be prevented.

Cloud troubleshooting

Depending on the technology being used, questions on this topic can be broad. In general, I stick with **Amazon Web Services Elastic Compute Cloud (AWS EC2)** or virtual machine or serverless troubleshooting. Typically, start with a description of the issue; details are often important in this type of question, so be attentive.

As an SRE, the first thing I'm looking for is how fast you go to the tools in the system, such as metrics or logs, to discover possible issues. For example, is the CPU at 100%? Does the serverless Lambda keep erroring out? Beyond just metrics, looking at security and permissions can also give good insight, but those issues can often be seen in the metrics.

Finally, I want to see that you're aware of some of the common shortfalls of specific technologies. I don't mean in terms of making me a list, but if you can't reach a virtual server, is a firewall at play causing an issue?

Load balancing and computer failure

Intermittent errors or failure, which only happen sometimes, are some of the most difficult to catch; when this is happening in systems with a load-balancing architecture, when compute instances fail, it can cause some, but not all, API calls to fail.

As this is an SRE interview question, there are a few different items at play here, not just the failed compute instance, which is likely to need replacing or rebooting. Understanding how intermittent failures can impact downstream systems is imperative. Knowing that a malfunctioning computer can still be attached to load balancing and how that may or may not automatically be replaced is essential in getting the system back online.

Finally, understanding how the load balancer and computer interact with downstream systems with retry capabilities and the detailed settings possible inside a load balancer to help minimize downtime are both areas of bonus points for this question for me.

Large-scale problems with multiple parts

When problems happen in a large-scale deployed application, we often find ourselves faced with a mess of databases, compute, and other miscellaneous architecture. Understanding and being able to find the needle in this haystack is key to being an SRE.

For me, this question is about the technical skill required to understand how the systems work together, where the weak points might be, and most importantly, how you break down and not become overwhelmed in large-scale applications.

Your best constant improvement tale

Improvement is at the center of all things to do with being an SRE, so asking about a time when you made constant improvements to a system is at the heart of reliability.

In the answer, it's more than just making a single change; it should be a story of how, over time, you continued to bring improvement to a system – a journey, not a single act. It should include the people side of the equation too and how you worked with others to bring about the change.

Bonus points for me are bringing in products or other teams to help them understand the improvement or, better, having them be part of defining the change.

Relating a system symptom with a cause

Asking about an issue is quite different from defining the cause. Take, for example, a service that won't respond to HTTP requests; the cause is quite simple and the number of possible failures is long.

We want to see you bring together the failure with other items such as metrics or logs to determine directionality in the resolution instead of wildly poking with a sharp stick to try to find the problem. We want to see purpose in the actions you take.

This certainly does not cover the gambit of technical questions, but it's a common list you'll see often and there are many variations. Remember, the interviewer won't classify these questions as we have here – it's about understanding why they ask the questions they do, to give you an idea of the broad nature of the response that is appropriate.

Non-technical questions

Being a rockstar SRE isn't just about technical skill set; we want to understand how you work with other teams, co-workers, and leadership. These questions are specific to being an SRE, though you should already be familiar with the standard: *tell me about a time when...* questions are common it seems to all interviews these days.

How do you approach

How do you approach these questions? More than the answer being technically sound, we want to hear about the approach. I don't just want to know that you fixed something, I want to know how.

Problems

When working through problems, do you have a plan, a direction? How does your mind work to both identify and disqualify issues as the problem? For this, we want to see the mind going through countless ideals and thoughts, but in a systematic way.

Conflict

Conflict-like problem-solving can get complex rather quickly. It's not uncommon, being the ones responsible for watching and resolving issues in production, for us to find faults in the work or processes. Most people don't like being told they are wrong, and some will fight tooth and nail to deny fault.

We are looking for the tactics you use to identify the issue, bring it to others, and work through the resolution. Most importantly, I want to see as little blame and friction as possible in the answer.

Leadership gaps

There has never been leadership that is perfect; in that regard, we want to know how you work through issues with your leadership. It could be misunderstandings, conflict, or even simply helping them understand a new path.

What we look for in this answer is how you interact with leadership. How does that look with regard to other leaders and co-workers? Did you discuss the issue with co-workers first to bring a common consensus to leadership? Are you willing to accept an answer that you don't understand (or are you unwilling)?

Co-worker gaps

Working with others is a highly prized sentiment that some fail from time to time. Like conflict resolution, it's important to work to get along with your co-workers. But gaps in co-workers' knowledge and ability can cost time and confusion in the daily work life of a team.

We want to see how you work with the others on your team, not only to assist with questions but also how we identify those times when we need assistance as well. A bonus in this question is bringing into the answer a discussion of strengths, that everyone on the team will have different strengths, and how a team leverages the strengths of all.

On-call experience

Nothing difficult here: on-call experience may be important to the team, and I prefer bringing it up in the initial interview to ensure everyone understands the nature of the job they are applying for. Because it's often an intrusion into a person's personal time, asking about on-call experience, including asking about a time you missed an outage, is imperative.

Looking through this answer, we want to see both a dedication to the team and company and also a commitment to life outside the company – demonstrating work-life balance.

Next, we'll dive into my favorite type of questions in an interview, the odd ones that may make no sense to you but are surprisingly revealing.

Insightfully odd questions

Breaking up the interview with what seems like meaningless questions can bring an element of refreshment to an interview. Here are some of our favorite odd-question topics, and the answers we're looking for.

Single color dilemma

"If you paint all the cars in the world yellow, give me three things that would happen."

Silly? Absolutely. Are you thinking about different ways to differentiate cars now? Shades of yellow perhaps? Confusion with taxis tends to come up a lot too. Here are the two things we're looking for.

First, what is your response to a question you certainly never heard before, is remarkably not what you expected, and hits you by surprise? How do you handle it? Did you take it in your stride, or did it cause you stress? Since SREs are often playing a firefighting role, seeing them react in real time is imperative.

Second, and this is so simple – did you give us three answers, or did you give up? I won't hire you if you don't give me three answers, period. I want SREs who are capable of working through issues and not giving up at the first sign of the question being too hard.

Seeking your own answers

These questions are typically about searching out your own answers. One of my favorites is how would you determine how many vacation days you can roll over into the next year?

Simple questions like this show us how you go about finding your own answers. For this question, the best answer is found in an employee handbook or online resource for the company – and not by asking others. In fact, asking others is the least reliable way of finding this answer.

Do I want an engineer who has to ask for help with simple questions? Absolutely not. We want to see independent thought of where to go for these answers.

Part of the interview should also be about what you want in a company, is this a good match, will I be happy with this career choice – a topic we cover next in the chapter.

What should you look for in a career?

A job is often just some place you work till you find direction or meaning in a position that gives you more than just a paycheck, and this is what we call a career. We'll walk through what you may want to review in making a career choice.

Define a good boss

They say people don't leave their jobs, they leave their bosses. And yes, I am very familiar with leaving due to leadership, which I've done multiple times in my career journey.

Tech compatibility

Understanding what your leadership thinks about the technologies you use is an important part of being a rockstar SRE. Beyond the monumental task of moving entire dev teams and others in product and development leadership, if your own leadership isn't versed in simple SRE concepts such as continuous improvement, it makes your job that much more difficult.

Blowing smoke and passing out cookies

Most people like to be recognized and made to feel special. But when leadership is handing out positivity for every small win, it gets hard to understand what's actually a win, and it starts to lessen the impact of the *thank you*. It's even worse when the improvements needed aren't pointed out.

I appreciate a good job, a thank you, and a way to go, but I'm not a fan of not providing feedback. My football coach used to say it's okay to be happy, it's not okay to be satisfied – meaning, it's great to win, but it's not okay to think the win is the only thing – we can always do better.

Honesty

Being a very honest and open person, I enjoy this quality in my leadership. Not every manager can say the truth to you; instead, they always try to sugarcoat or, worse, never give you feedback. It's something you'll want to be aware of; levels of honesty with your leadership should be reviewed as they will impact your career.

Are they in control?

Yes, there is leadership that is not in control. This can range from being unable to make decisions to being out of sync with higher levels of leadership. Now, a manager isn't always going to walk into a room and bark orders like a drill sergeant, but those unwilling to engage can be problematic, especially in a role where, inevitably, they will need to step in on your behalf to encourage others to engage.

Planning

Is there a plan? For the team? For the products? For the company? Understanding the direction and that it's a common thread through multiple teams is imperative to a business's success. If teams are all doing whatever they want, this could be an issue – especially when you are tasked with working directly with them.

Dotted line reporting

There is nothing wrong with having a phantom manager, and it's actually quite common. It can be advantageous in many ways, but you need to ensure the manager who is in charge of you from an HR perspective is okay with this because, if not, the conflict will not be fun and can cost you your career.

Morals

Is this the right thing to do? It's an important question, and some companies are willing to trick customers and even employees. This can be as simple as naming a new fee on customer bills so it looks like a new federal or state tax or even misrepresenting the company or products.

We have to decide whether we are okay with the morality of the company – and nobody can make that call but you. For those who have strong morals, this is something that will be difficult to swallow in a job.

We've been diving into the employee relationship with a company as it impacts your career.

Next, we'll dive into the research you should do into the company itself.

Researching the company

Being part of a good company can be exciting and fulfilling in so many ways. I love companies that will stand behind their employees and provide the resources they need to be better. Understanding the business model and its stability for the next decade will ensure you are not stressed about losing your job.

Business model

A company should make money (understanding how can be complex and not entirely transparent), but it is key to identifying the strength of the company. Take, for example, social media sites, where the business model is to use content created by subscribers to bring in and keep subscribers on the site to sell advertising. This business model is designed to sell advertising, and it's the advertisers, not the visitors, who are the customers.

Profitability for the next decade

When you look at the business model, is it sustainable? What happens if the economy changes? Could a new product destroy sales?

These are all relevant questions and if you have questions about this, you should ask the company; a strong company won't have a problem sharing its strengths.

Structure

The structure of the company can play into how leadership works within it. From a tech perspective, when we look at the structure, there are a few common ideologies out there. One is that a single corporate leader owns both product and technology, and while I've seen this work, when there is a separate owner of technology reporting to the CEO or president, the healthy tension it creates can instill better leadership and a set of checks and balances for product and technology alike. The same is true of security being separated from technology.

In general, my preference is for the product to be independent of technology, though the teams should work very closely.

Large versus small

There are advantages to working for small and large companies alike. It's more of a personal preference, but it should absolutely be considered when looking at a company. It should also be noted that it's easier to shine in a small company, though it can come at the cost of a higher workload and skewed work-life balance.

Public versus private

This, too, is a preference. Public companies often have much more routine in their training, policies, and structure than private companies. From audits to policy, there can be a very large difference.

If you have never worked for a public company, the goals can be a tad different as there is an overall board who must answer to the stockholders; this can cause some leadership decisions and processes, such as approving annual raises, to take a longer period of time.

Online reviews

Finally, there is always a plethora of information on the internet about companies. Like all things online, this should be taken with a word of caution. For example, companies with large call centers often have lower reviews because call center positions tend to have lower wages and higher turnover, which can draw down the ratings of a company overall. In those situations, the technology side of the company could be a dream to work for, but you can't tell from looking at the raw rating for the entire company.

It's also important to look at the age of the reviews; given how many of these sites have been around for more than five years, stale data can cause issues in ratings in companies where real reform done by new leadership has taken hold and lifted the entire company.

Overall, understanding the company you want to work for is important; it's also important to understand whether it fits into your overall career plan. Are you okay with a short 2-3-year stint at a company to gain experience and move on, or are you looking for a longer-term commitment? The idea is to find your fit and be conscious of your decision.

As part of your career journey, your credentials can play a significant part, from degrees to certifications. In this next section, we'll discuss how certifications play into hiring.

Are you over-or under-certified?

Certifications can be a great plus on a resume and help you get the interview. In short, they are good things; however, like any good thing, there is a point where they can become negative. In the industry today, it seems everyone is offering certifications, but only a few have real value in the world of hiring.

Certifications that matter

So, what certifications matter? This is a rather simple answer – cloud provider certifications and a few vendor certifications, including Terraform and Kubernetes. The list is rather small as to what certifications will actually add value to your resume; for me, the only ones I list are my **Amazon Web Services (AWS)** professional certifications, leaving out lesser certifications.

How many are too many certifications?

I once held the resume of a person who was three years into an IT career and had every AWS certification and every Azure certification available listed on their resume. I refused to even glance at the rest of the resume because of this. You see, in three years, you would barely be able to pass both professional certifications for AWS, and certainly, you wouldn't have the skill set for the Azure certification as well in that time. We obtain certifications by studying, and while there is nothing wrong with that – in fact, I always study for my certifications – not having the real-world knowledge behind them is problematic when used as a merit for hiring.

Relevancy

Like old Walkman portable audio tape players, technology gets replaced with newer versions. This is why the CompTIA A+ certification I received in 1998 isn't on my resume – first, it's far too simple given my ability level, but also, the technology is ancient by any standards.

The same is true of entry-level and associate certifications. While it's okay for a product owner to state they have the very basic AWS Cloud Practitioner certification, which provides a basic overview of the AWS products, it would actually hurt my resume to list it, as it's too basic and too simple for my other skill sets.

When listing your certifications, make sure they speak to both your position and level of knowledge. While some may think listing everything is the right way to go when scanning resumes, sometimes, too much irrelevancy is a turn-off.

Finally, this last section of the chapter will go over a few interview tips and salary negotiation techniques that have been very good to me in the past while negotiating a rate.

Tips for landing the job with a great salary

So, the day of the interview is here, and you're both excited and a little nervous! It happens to most of us, and it's completely normal. We'll talk through some great interview tips, salary negotiations, and finally, walk through transiting to a new job, including talking about some of the interviewing tips that have given me some of my greatest yeses during my job searches in the past.

Interview tips

Interviewing is a skill that is learned over time, so if you have a chance to interview for a job, you should. Building this skill, and admittedly, keeping the skill sharp by doing an interview or two a year, will build this skill like a muscle.

The basics

Be on time, but not too early. And yes, I've arrived an hour before an interview, so I could ensure I found the company location, drove around the loop, and ensured I knew where the entrance and parking were. In general, you should show up no more than 15 minutes before the interview, with 5 to 10 minutes being the sweet spot.

If your meeting is online, you should try the bridge, test your audio and video connections before the interview, and show up five minutes early to the bridge. I recommend that if you haven't used the online meeting platform they use, you do a test hours before the interview. Poor or non-working audio/video is an instant no in my book for an SRE.

Mind your space and manners; this includes ensuring you are not disturbed, nothing is going on behind you, and you are polite. Oh, and don't smoke during your interview – I can't believe I have to say that, but it's happened. If you are time constrained or have to call in from a car or a noisy area, make sure the interviewer knows and agrees. I once interviewed in a quiet corner of an airport, but I was sure to ensure it would be okay first.

Finally, remember to be communicative, before and after. Nothing is more polite than an excited email about an upcoming email or a thankful response after the interview from the candidate – and please, make it personal and, if possible, a little funny if you think that fits with the culture.

Nervousness

This is huge in interviews, and I've seen it time and time again. One of the best engineers on my team now was nervous beyond belief when talking with my team. From a hiring manager's perspective, you'll want to lead the candidate into an area they feel very comfortable in, something they know and know well – or better yet, talk about something they are passionate about. And if you are on the other side, have a few ideas of fun things you've enjoyed doing, know like the back of your hand, and can speak fluently about written on a note in front of you, and go to those subjects when you get nervous.

Passion

Showing passion during an interview should be a number one priority, though, I would stop before you become a cheerleader version of yourself. Ensure you know what your passions are, learn something about the company, and talk about what you know.

Showing passion during an interview will make you stand out; it shows you as someone who has drive and will be interested in and have fun with their work – and therefore, more willing to put more into their work.

That same passion will make you appear more valuable because you are, and can be a plus in salary negotiations, which we'll dive into next.

Salary negotiations

This is never the fun part of the interview process. Ask for too much and they won't hire you; ask for too little, and you'll miss out. So, what's the trick here? Well, many states are now forcing companies to disclose salary ranges for their positions, though don't get too excited over the biggest number; remember, it's a range.

Where to begin? Simply ask what the budget range is for a candidate as qualified as you would be, then ask for a little bit more. Make sure to take into account benefits such as 401(k) matching and insurance, which can range from thousands to tens of thousands worth of benefits a year. Use an online salary calculator to estimate what your pay would be with insurance and other costs taken out every month.

Never give ultimatums in the first rounds of salary negotiations; it's a conversation. Be honest, open, and truthful; the last thing you want is to take a job and be unhappy with the pay.

Summary

Winning over the hearts and minds of a new employer takes some knack, knowledge, skill, and a bit of luck. I've nailed interviews and failed miserably. The number one rule is to get back up, dust yourself off, and keep trying.

When you start your search, be mindful of the places you want to work for and their culture. Make a list of things important to you. There is nothing wrong with taking a job to allow you to look for the perfect career – just never give up.

This book, for us, has been a journey based on our understanding of decades of experience in the technology field, a place where we have found passion and meaning in our lives. We want to thank you for taking the journey through this book and hope you do your best to find the same passion in your own lives!

Most importantly, good luck to you in your search and future career as a rockstar SRE!

Appendix A – The Site Reliability Engineer Manifesto

Often, we have received a question such as, "*What does a site reliability engineer do?*". Although we could list all site reliability engineering activities with some effort, how long would the list be valid or complete in a domain where the technology changes constantly?

Therefore, we don't define site reliability engineering by what a **site reliability engineer** (SRE) does but by which principles a whole organization follows and by the skills the SREs who work at these organizations develop.

In essence, they should be accomplishing the following goals:

- Follow through on the site reliability engineering **tenets**
- Implement site reliability engineering **best practices**
- Employ site reliability engineering **techniques**
- Develop site reliability engineering **skills**

However, we understand there are contractual language limitations and constraints on articulating such *loose definitions* as guiding principles. Therefore, we want to publicly express our opinion regarding an SRE's primary responsibilities without linking them to any specific technology, product, or platform. The best way to convey this message is through a manifesto of SREs, by SREs, for SREs. This appendix is the result of some discussions among ourselves regarding these responsibilities. The latest version is in the **GitHub** repository at this link: `https://github.com/PacktPublishing/Becoming-a-Rockstar-SRE/blob/main/Appendix/the-sre-manifesto.md`.

The manifesto

We offer the first version of *The Site Reliability Engineer Manifesto* here.

We, the *Becoming a Rockstar SRE* authors and technical reviewers, met together to discover better ways of defining the **SRE** persona.

We believe we can create a more unified, concise, and solid SRE persona by stating that an SRE's essential responsibilities are as follows:

- SREs make sure their systems are reliable and not just available and resilient
- SREs guarantee all systems and applications are observable and under monitoring
- SREs manage systems, services, and infrastructure to learn how to automate arduous work
- SREs use data science and statistical methods to understand the observability data
- SREs identify, measure, and reduce arduous work arising from operations and engineering
- SREs implement test cases, execute software delivery tests, and stay ahead with capacity planning
- SREs employ chaos engineering to unveil systemic weaknesses in production

We welcome anyone to use this manifesto at will, as we have made it public. We will review this manifesto in the GitHub repository and update it yearly.

How to adopt it

Please copy and paste it to your organization's internal portal. Of course, you will need to explain each item to your SREs and constantly check whether they are embracing these responsibilities. Remember to ensure they align with your company's values and culture.

How to contribute to it

If you think another key SRE responsibility should be part of *The Site Reliability Engineer Manifesto*, please create a *pull request* with your proposal in the **GitHub** repository.

We want to thank you in advance for any contributions. May the light of reliability always be over your head. As a famous SRE once said, live long and prosper.

Appendix B – The 12-Factor App Questionnaire

As we said in the beginning, site reliability engineers must be involved in all phases of the solution life cycle (see *Chapter 1, SRE Job Role – Activities and Responsibilities*), including the design phase. When cloud-native applications became a reality, intelligent people began defining good application designs to exploit all the cloud features and capacities. As we collectively learned, we construct reliable systems with well-designed applications by applying **site reliability engineering** (**SRE**) principles. One great work on enlightening ways to have more robust, resilient, and scalable application designs is, without a doubt, *The Twelve-Factor App* manifesto.

It was written by Adam Wiggins and translated into multiple idioms. In its original text, you can read this: "*This document synthesizes all of our experience and observations on a wide variety of software-as-a-service apps in the wild. It is a triangulation on ideal practices for app development, paying particular attention to the dynamics of the organic growth of an app over time, the dynamics of collaboration between developers working on the app's codebase, and avoiding the cost of software erosion.*" You can access the manifesto at this link: `https://12factor.net/`.

The 12-Factor App Questionnaire is derived from *The Twelve-Factor App* manifesto based on the many questions and answers about good app designs in the forums. The latest version is in the **GitHub** repository at this link:

```
https://github.com/PacktPublishing/Becoming-a-Rockstar-SRE/blob/
main/Appendix/the-12-factor-app-questionnaire.md
```

The questionnaire

We aligned this questionnaire with the 12 factors published by **Heroku**. The following sections hold questions for each dimension that can objectively evaluate an application design's resiliency and reliability. Ideally, the application design should follow these factors before writing any code. However, unfortunately, site reliability engineers are often involved later in the **software development life cycle** (**SDLC**), so they can utilize this questionnaire to pinpoint areas of improvement in the application domain.

In the following sections, we describe each factor, point to further details in the manifesto, explain why this is important from the SRE perspective, and give a grade on how this factor affects system reliability. We adopted a **Fibonacci**[1] sequence ranking:

- **Grade 1**: Minor impact on the reliability

- **Grade 2**: Low impact

- **Grade 3**: Low-medium impact

- **Grade 5**: Medium impact

- **Grade 8**: High impact

- **Grade 13**: Very high impact

Each question in the next tables (see *tables B.1* to *B.12*) is a closed question with two possible answers: yes or no. A favorable reply can be interpreted as compliant or non-compliant with the 12-factor app principle, depending on the question. We indicate which answer should be considered compliant and give examples when the response is *affirmative* for each question in the end.

> **Important note**
>
> Notice that for some questions, being compliant means answering yes to them. For others, you must answer no to respect the principle. We purposefully crafted this questionnaire structure to avoid a single answer for all questions without even reading the questions.

Let's get into the questionnaire, organized by the following dimensions.

Factor I – Code base

A single and unique code base is tracked in revision control, many deploys:

- **Description**: Application code is continuously tracked in a version control system. There is only one codebase per app, but many deployments will exist.

---

1. The Fibonacci sequence or series is a mathematical construct in which the following number in the sequence is the sum of the previous two numbers, starting with 0 and 1.

- **Details**: Codebase factor (`https://12factor.net/codebase`).

- **Why**: Having different code bases for the same app leads to technical debt, merging problems, dark code[2], and vulnerabilities.

- **System reliability impact**: 2 – LOW.

Questions

Item ID	Question	If answered YES	If answered NO	Affirmative answer examples
1.1	Is the application source code stored in a source versioning-controlled repository?	Compliant	Non-compliant	Git, Mercurial, or Subversion
1.2	Is the application source code stored in a single repository or multiple repositories that share a root commit?	Compliant	Non-compliant	GitHub, GitLab
1.3	Do all the deployments of this app share the same codebase?	Compliant	Non-compliant	GitHub Actions, Argo CD, CI/CD

Table B.1 – Factor I questions, compliance indication, and examples

Factor II – Dependencies

Explicitly declare and isolate dependencies:

- **Description**: The app declares all dependencies, completely and precisely, via a dependency declaration manifest. Furthermore, it uses a dependency isolation tool during execution to ensure that no implicit dependencies "*leak in*" from the surrounding system. The full and explicit dependency specification is applied uniformly to production and development.

- **Details**: Dependencies factor (`https://12factor.net/dependencies`).

- **Why**: Having implicit or system-wide dependencies leads to app disruptions if any changes happen to such dependencies.

- **System reliability impact**: 8 – HIGH.

---

2   Dark code is any code snippet inserted in software that is not traced or trackable. For instance, code amends are done outside the code base to fix an issue in the deployment.

Questions

Item ID	Question	If answered YES	If answered NO	Affirmative answer example
2.1	Does your code refer to or use any external components considered hardcoded?	Non-compliant	Compliant	System DLLs, `GetTempPath`, Windows services, Windows registry, `ConfigurationManager`
2.2	Does your code perform any operations that create an external dependency?	Non-compliant	Compliant	Using filesystem, directory manipulation, file manipulation
2.3	Does your code use a dependency declaration method?	Compliant	Non-compliant	Pip (Python), npm (Node.js), Virtualenv (Python), Fatjar (Java EE – JEE), Project Object Model (POM) XML (JEE)

Table B.2 – Factor II questions, compliance indication, and examples

Factor III – Config (configuration)

Store config in the environment:

- **Description**: The app stores config (configuration) in environment variables (often shortened to **env vars** or **env**). Env vars are easy to change between deploys without changing any code.

- **Details**: Config factor (`https://12factor.net/config`).

- **Why**: Storing application config inside the code leads to scalability constraints and security issues.

- **System reliability impact: 8 – HIGH.**

Questions

Item ID	Question	If answered YES	If answered NO	Affirmative answer example
3.1	Does your code store config (credentials, IP addresses, API tokens, hostnames) as constants?	Non-compliant	Compliant	Username and password
3.2	Does your app strictly separate config from code?	Compliant	Non-compliant	No hardcoded settings
3.3	Does your app store config in environment variables?	Compliant	Non-compliant	`export DB_URL="https:/..."`

Table B.3 – Factor III questions, compliance indication, and examples

Factor IV – Backing (backend) services

Treat backing services as attached resources:

- **Description**: The application makes no distinction between local and third-party backing services.
- **Details**: Backing (backend) services factor (`https://12factor.net/backing-services`).
- **Why**: Resources can be attached to and detached from deploys at will. If a database resource fails, the app can switch to another database resource without changing the code.
- **System reliability impact**: 5 – MEDIUM.

Questions

Item ID	Question	If answered YES	If answered NO	Affirmative answer example
4.1	Does the app access backing (backend) services as attached resources via a URL or other locator?	Compliant	Non-compliant	Using any cloud-based services
4.2	Does the app treat any on-premises backing services as attached resources?	Compliant	Non-compliant	Company SMTP[3] server: `smtp://auth@host/`

Table B.4 – Factor IV questions, compliance indication, and examples

Factor V – Build, release, run

Strictly separate build and run stages:

- **Description**: The application strictly separates the build, release, and run stages.

- **Details**: Build, release, run factor (`https://12factor.net/build-release-run`).

- **Why**: Changes to the application in runtime are not restricted and only happen with a new release. The application can be easily rolled back to a previous release.

- **System reliability impact**: 8 – HIGH.

Questions

Item ID	Question	If answered YES	If answered NO	Affirmative answer example
5.1	Are the app releases changed or tampered with after it has been created?	Non-compliant	Compliant	Configuration drift
5.2	Can the app, deployed in runtime, be changed without a new build or release?	Non-compliant	Compliant	Unprotected Kubernetes cluster
5.3	Are you using an automated deployment platform?	Compliant	Non-compliant	Argo CD

---

3   SMTP stands for Simple Mail Transfer Protocol.

5.4	Does the current deployment orchestrator have automated rollback capabilities?	Compliant	Non-compliant	Argo Rollouts

Table B.5 – Factor V questions, compliance indication, and examples

Factor VI – Processes

Execute the app as one or more stateless processes:

- **Description**: App processes are stateless and share nothing among them.
- **Details**: Processes factor (`https://12factor.net/processes`).
- **Why**: The app takes full advantage of scalable cloud infrastructure. Service disruptions are graceful and smooth.
- **System reliability impact**: 3 – LOW-MEDIUM.

Questions

Item ID	Question	If answered YES	If answered NO	Affirmative answer example
6.1	Does the app use sticky sessions to cache session data in memory?	Non-compliant	Compliant	Using stateful session (Servlet/Spring), webform authentication
6.2	Does the app use microservices, event-driven, or a combined architecture pattern?	Compliant	Non-compliant	OpenAPI, Apache Kafka

Table B.6 – Factor VI questions, compliance indication, and examples

Factor VII – Port binding

Export services via port binding:

- **Description**: The app is entirely self-contained and does not rely on the runtime injection of a web server into the execution environment to create a web-facing service.
- **Details**: Port binding factor (`https://12factor.net/port-binding`).
- **Why**: The app doesn't rely on web servers or app servers.
- **System reliability impact**: 2 – LOW.

Questions

Item ID	Question	If answered YES	If answered NO	Affirmative answer example
7.1	Is the app self-contained and not dependent on any web or app server?	Compliant	Non-compliant	Using Express (Node.js) or Flask (Python)

Table B.7 – Factor VII questions, compliance indication, and examples

Factor VIII – Concurrency

Scale out via the process model:

- **Description**: Application processes should never spawn daemons (*daemonize*) or write **process ID (PID)** files. Instead, rely on the operating system's process manager.
- **Details**: Concurrency factor (`https://12factor.net/concurrency`).
- **Why**: The app is scalable horizontally as well.
- **System reliability impact: 5 – MEDIUM.**

Questions

Item ID	Question	If answered YES	If answered NO	Affirmative answer example
8.1	Does this app scale out through the native process model if there's more demand?	Compliant	Non-compliant	Multithreaded, parallel processing
8.2	Does this app have separate and distinct processes assigned to handle different requests and tasks?	Compliant	Non-compliant	One microservice for each type of request

Table B.8 – Factor VIII questions, compliance indication, and examples

Factor IX – Disposability

Maximize robustness with fast startup and graceful shutdown:

- **Description:** The application's processes are disposable and can be started or stopped immediately.

- **Details:** Disposability factor (`https://12factor.net/disposability`).

- **Why:** A high-reliability app has disposable processes, so it takes seconds to boot after a shutdown.

- **System reliability impact: 8 – HIGH.**

Questions

Item ID	Question	If answered YES	If answered NO	Affirmative answer example
9.1	Does this app finish processing all active requests before shutdown when it receives a termination signal (`SIGTERM`)?	Compliant	Non-compliant	Two-phase commits in transactions
9.2	Does the app take over 1 minute to reboot after a shutdown?	Non-compliant	Compliant	N/A

Table B.9 – Factor IX questions, compliance indication, and examples

Factor X – Development/production (dev/prod) parity

Keep development, staging, and production as similar as possible:

- **Description:** The application is designed for continuous deployment by keeping the gap between development and production small.

- **Details:** Dev/prod parity factor (`https://12factor.net/dev-prod-parity`).

- **Why:** Differences between backing (backend) services mean tiny incompatibilities crop up, causing code that worked and passed tests in development or staging to fail in production.

- **System reliability impact: 5 – MEDIUM.**

Questions

Item ID	Question	If answered YES	If answered NO	Affirmative answer example
10.1	Does the app synchronize its development, staging, and production environments daily?	Compliant	Non-compliant	Similar infrastructure and configuration to proactively capture problems
10.2	Does the app use the same dataset or other backing services in all environments?	Compliant	Non-compliant	Common datasets and services for meaningful testing

Table B.10 – Factor X questions, compliance indication, and examples

Factor XI – Logs

Treat logs as event streams:

- **Description**: The app never concerns itself with routing or storing its output stream.

- **Details**: Logs factor (`https://12factor.net/logs`).

- **Why**: A critical part of any system observability is the logs. Reliable apps need to produce logging without affecting their functionalities.

- **System reliability impact**: 13 – VERY HIGH.

Questions

Item ID	Question	If answered YES	If answered NO	Affirmative answer example
11.1	Does the app treat logs as streams?	Compliant	Non-compliant	Using `log4j`
11.2	Does the app have code that manages specific log files?	Non-compliant	Compliant	The app handles log storage and management
11.3	Is app logging managed by a log management tool to store data separately instead of creating code that addresses the log?	Compliant	Non-compliant	Using Grafana Loki

Table B.11 – Factor XI questions, compliance indication, and examples

Factor XII – Admin processes

Run admin/management tasks as one-off processes:

- **Description**: One-off (nonrepeatable) admin processes such as installation should be run in an identical environment to the app's regular long-running processes.

- **Details**: Admin processes factor (`https://12factor.net/admin-processes`).

- **Why**: One-off (non-repeatable) admin processes run against a release, using the same code base and configuration as any process run against that release. That approach minimizes errors and inconsistencies.

- **System reliability impact**: 8 – HIGH.

Questions

Item ID	Question	If answered YES	If answered NO	Affirmative answer example
12.1	Does the app ship administration code with its core code?	Compliant	Non-compliant	Build-to-manage principle
12.2	Do the app's one-off admin tasks run in the same (or similar) environment as its long-running regular processes?	Compliant	Non-compliant	REPL[4] console

Table B.12 – Factor XII questions, compliance indication, and examples

How to adopt this questionnaire

Please copy and paste it to your organization's internal portal, then use it to assess your existing applications. Based on the non-compliant answers, the SRE team can create a list of action items to improve an application's design and work with developers to refactor the code. After some iterations, you can even add more questions to *The 12-Factor App Questionnaire*.

How to contribute to this questionnaire

If you think another key question should be part of *The 12-Factor App Questionnaire*, please create a pull request with your proposal in the **GitHub** repository.

We appreciate any contributions from the SRE community. May the reliability winds bless your systems, but remember: "*hope is not a strategy*".

---

4. The **read-eval-print-loop** (**REPL**) console is a CLI that prints out a prompt, reads an input command, evaluates it, then prints the result repeatedly.

Index

Packtpub.com

Subscribe to our online digital library for full access to over 7,000 books and videos, as well as industry leading tools to help you plan your personal development and advance your career. For more information, please visit our website.

Why subscribe?

- Spend less time learning and more time coding with practical eBooks and Videos from over 4,000 industry professionals

- Improve your learning with Skill Plans built especially for you

- Get a free eBook or video every month

- Fully searchable for easy access to vital information

- Copy and paste, print, and bookmark content

Did you know that Packt offers eBook versions of every book published, with PDF and ePub files available? You can upgrade to the eBook version at packtpub.com and as a print book customer, you are entitled to a discount on the eBook copy. Get in touch with us at customercare@packtpub.com for more details.

At www.packtpub.com, you can also read a collection of free technical articles, sign up for a range of free newsletters, and receive exclusive discounts and offers on Packt books and eBooks.

Other Books You May Enjoy

If you enjoyed this book, you may be interested in these other books by Packt:

Cloud-Native Observability with OpenTelemetry

Alex Boten

ISBN: 9781801077705

- Understand the core concepts of OpenTelemetry
- Explore concepts in distributed tracing, metrics, and logging
- Discover the APIs and SDKs necessary to instrument an application using OpenTelemetry
- Explore what auto-instrumentation is and how it can help accelerate application instrumentation
- Configure and deploy the OpenTelemetry Collector
- Get to grips with how different open-source backends can be used to analyze telemetry data
- Understand how to correlate telemetry in common scenarios to get to the root cause of a problem

SAFe® for DevOps Practitioners

Robert Wen

ISBN: 9781803231426

- Understand the important elements of the CALMR approach

- Discover how to organize around value using value stream mapping

- Measure your value stream using value stream metrics

- Improve your value stream with continuous learning

- Use continuous exploration to design high-quality and secure features

- Prevent rework and build in quality using continuous integration

- Automate delivery with continuous deployment

- Measure successful outcomes with Release on Demand

Packt is searching for authors like you

If you're interested in becoming an author for Packt, please visit `authors.packtpub.com` and apply today. We have worked with thousands of developers and tech professionals, just like you, to help them share their insight with the global tech community. You can make a general application, apply for a specific hot topic that we are recruiting an author for, or submit your own idea.

Share your thoughts

Now you've finished *Becoming a Rockstar SRE*, we'd love to hear your thoughts! Scan the QR code below to go straight to the Amazon review page for this book and share your feedback or leave a review on the site that you purchased it from.

https://packt.link/r/1803239220

Your review is important to us and the tech community and will help us make sure we're delivering excellent quality content.

Download a free PDF copy of this book

Thanks for purchasing this book!

Do you like to read on the go but are unable to carry your print books everywhere?

Is your eBook purchase not compatible with the device of your choice?

Don't worry, now with every Packt book you get a DRM-free PDF version of that book at no cost.

Read anywhere, any place, on any device. Search, copy, and paste code from your favorite technical books directly into your application.

The perks don't stop there, you can get exclusive access to discounts, newsletters, and great free content in your inbox daily

Follow these simple steps to get the benefits:

1. Scan the QR code or visit the link below

https://packt.link/free-ebook/9781803239224

2. Submit your proof of purchase
3. That's it! We'll send your free PDF and other benefits to your email directly